SUPERCARS

The Story of the Dodge Charger Daytona and Plymouth SuperBird

By Frank Moriarty

Foreword by Buddy Baker

Featuring Photographs from the Pal Parker Archives

Designed by Portia Moriarty
Edited by Ross A. Howell, Jr. and Katherine A. Neale

©1995 by Frank Moriarty. All rights reserved.
This book, or any portion thereof, may not be reproduced or transmitted in any form or by any means, electronic or mechanical, including photocopying, recording, or by any information storage and retrieval system, without permission in writing from the copyright holder.

Printed in Hong Kong
Published by Howell Press, Inc., 1147 River Road, Suite 2, Charlottesville, Virginia 22901.
Telephone (804) 977-4006
First Printing

Library of Congress Cataloging–in–Publication Data
Moriarty, Frank.
    Supercars: the story of the Dodge Charger Daytona and Plymouth SuperBird / Frank Moriarty p. cm.
    ISBN: 1-57427-043-5

1. Stock car racing —United States—History. 2. Dodge Charger automobile—History. 3. Plymouth Automobile—History. I. Title.

GV1029.9.S74M67 1995   796.7'2'0973
                                    QBI95–20204

The Publisher recognizes that some words, model names, and designations, for example, mentioned herein are the property of the trademark holder. We use them for identification purposes only. This is not an official publication. Every effort has been made by the publisher to ensure permission has been obtained and proper credit is given regarding the use of any photographs or illustrations which appear in this publication.

Illustration/Photo Credits

Chrysler Historical Collection — 29, 48, 85, 88, 89, 93 (top), 97
Dave Ferro — 132
Larry Gray — 141 (right)
Garry Hill — 56
Owen Kearns, Jr. Collection — 129
Tim Kirkpatrick Collection — 87 (top), 137 (top), 140 (bottom), 148 (right)
Dick Lajoie Collection — 8, 12, 13, 14, 16, 18, 20, 21, 51, 52, 87 (lower left), 109, 110, 111, 116, 117, 119, 120, 121, 122, 123
McCaig-Wellborn International Motorsports Research Library — 32, 33, 41 (right), 43, 84, 91, 95
Steve Mirabelli — 148 (left), 149 (bottom), 150-160
Frank Moriarty — 7, 11 (left), 23, 26, 35, 54, 55, 82, 86 (lower left), 90, 113, 114, 118, 126, 130, 131, 134-135, 137 (bottom), 138, 139, 140 (top), 141 (left), 142 (top left), 143, 144, 145, 146-147
Portia Moriarty — 142 (top right, bottom)
Pal Parker Archives — 30, 39, 41 (left), 42, 44, 45, 58, 60, 61, 62-63, 64, 65, 66, 67, 68, 69, 70, 72, 75, 76, 77, 79, 81, 127
Larry Rathgeb Collection — 5, 11 (right), 15, 22, 25, 28, 36, 40, 46, 47, 59, 71, 86 (lower right), 87 (lower right), 92, 93 (bottom), 98, 99, 100, 101, 102-103, 104, 105, 106, 107, 108, 115, 124, 128
Doug Schellinger Collection — 86 (top), 125
Don Snyder Collection — 94, 149 (top)
Illustrations on pages 17 and 19 reprinted with permission from SAE paper #700036 © 1970 Society of Automotive Engineers, Inc.

Contents

Acknowledgments   4

Foreword by Buddy Baker   5

Introduction   6

Muscle Mania and the Need for Speed   8

How Chargers Grow Wings   16

The Summer of '69   30

Building a Better Bird   48

Aero Wars   56

Wild in the Streets   82

Where No Man Has Gone Before   98

The End of an Era and Wings that Never Were   116

The Winged Legacy   130

Appendix   148

Acknowledgments

This book would not have been possible without my wife Portia, who provided ideas, support, and advice. She not only listened to countless tales of woe during the years this text was completed, she also designed this book. To Portia, many thanks with much love.

Ross Howell at Howell Press was excited at the prospect of a book about the Daytonas and SuperBirds as soon as he heard about the idea, and I thank him for his interest in making the idea a reality. Judith Gotwald at Creation worked with Portia in the design of this book, and the results are appreciated.

Thanks to my pal, master lensman Rick Farnkopf, for his photography advice and assistance.

All of the Daytona and SuperBird owners and fans I talked with demonstrated great interest and support, but in particular I must express special gratitude to the car owners who made special arrangements so I could photograph their beautiful vehicles — Bill Stech in New Jersey, Tim Kirkpatrick in Pennsylvania, John Pappas in Michigan, and Don Snyder in New York. Your cooperation is greatly appreciated.

In Talladega, Alabama, Betty Carlan was a tremendous help at the McCaig-Wellborn International Motorsports Research Library with historical materials, suggested contacts, reminiscences, and plenty of kind hospitality — thank you, Betty.

Tim Wellborn made it possible for me to photograph the number 88 and 71 Daytonas, as well as his own beautiful SuperBird, at the International Motorsports Hall of Fame and Museum at Talladega. Tim also provided much insight, advice, and a look at the inner workings of the 1994 Aero Warrior Reunion, and I thank him for fielding so many phone calls and answering so many questions.

Brandt Rosenbusch at the Chrysler Historical Collection assisted me with photography and contacts, and made it possible for me to reach Gary Romberg, who opened many doors.

Gary kindly arranged a meeting at Chrysler's amazing Technical Center where he, John Pointer, and Dick Lajoie provided a colorful overview of the entire wing car development process. Dick Lajoie graciously loaned me many materials from his collection, and Gary was a help several times with follow-up questions. Thanks to you all.

Many drivers, car owners, and people connected with wing cars, the sport of stock car racing, or Chrysler Corporation contributed their comments or assistance to this book. Thanks to: Pete Hamilton, Richard Petty, Martha Jane Bonkemeyer at Petty Enterprises, Jim Hunter, Stan Creekmore, Dick Hutcherson, Jim Vandiver, Geoff Bodine, Verne Koppin, Don Mirzaian, Buddy Baker, Dick Brooks, John Herlitz, Bob Marcell, George Wallace, Dave Ferro, Harry Hyde, Cale Yarborough, Charlie Glotzbach, Richard Brickhouse, Bud Moore, Greg Moore, Bill Brodrick, Owen Kearns, Jr., Frank Wylie, Larry Gray, Eddie Gossage at Charlotte Motor Speedway, Al Robinson at Dover Downs International Speedway, and Jim Freeman at Talladega Superspeedway.

Ron Drager at Automobile Racing Club of America and Donald Davidson at United States Auto Club helped research wing car race performances in their sanctioned events, and Doug Schellinger of the Daytona-SuperBird Auto Club was also a great help. Greg Fielden's invaluable reference work *Forty Years of Stock Car Racing* should be in the collection of every race fan, as it is truly stock car racing's bible. Thanks also to the staff at the National Automotive History Collection of the Detroit Public Library, as well as Louis Helverson at the Automobile Reference Collection of The Free Library of Philadelphia for assistance with race results and vehicle production information.

Thanks to Pal Parker, whose photographs from that great era of NASCAR truly capture the wing car excitement, and motorsports artists Steve Mirabelli and Garry Hill. Portions of Steve's "Wing Warriors" print are reproduced in this book, but the entire print is available from Car Stars Enterprises, 315 Miller Ave., Trenton, NJ 08610. Information on Garry's print of the historic 1970 Daytona 500 wing cars is available from Garry Hill Automotive Fine Art, P.O. Box 1311, Mooresville, NC 28115-1311. The Pal Parker Archives may be reached at P.O. Box 218, Flagler Beach, FL 32136.

Finally, I must offer special thanks to Larry Rathgeb. In his work at Chrysler's Special Vehicles Group, Larry oversaw the development of the wing cars and their amazing race performance. Larry's generosity with his time and the materials in his collection have largely made this book possible, and I am indebted to his kindness. Thank you, Larry, for your interest in this book and for helping bring the world two amazing automobiles.

Foreword

On March 24, 1970, Buddy Baker — winner of 19 races in NASCAR's top stock car series — roared into history at the wheel of a Dodge Charger Daytona when he became the first person to officially drive a speedway lap at over 200 MPH.

In the late 1960s NASCAR had just left the weekly shows, running on half-mile dirt tracks and races like that. We really didn't progress into the major speedways until just about the time the Hemi cars came about. The minute the Hemi got on the racetrack, it was awesome. Everything else, every other car that had been run before was kind of obsolete compared to the Hemi. And then came the wing cars.

I always did a lot of testing work so there's a lot of things I've been involved in that didn't show up as far as my stats, but I feel pretty proud of them. One of those things is that I was in on the original testing of the wing cars.

Up until then, race cars were built on the tank theory — make it strong and make it durable and don't worry about it being real sophisticated as far as handling. But the first time I looked at the Daytona I just thought, "Wow!" That was the first body I had ever seen that had been designed to be a race car. There wasn't any question about what that thing was made for — it was made to eat Ford's Cobras. The Daytona was the first mongoose, and everything on that car was made for a reason. It was the first truly aerodynamic car, period.

At the 25th anniversary celebration of the wing car at Talladega, I had just gotten up to the microphone and I said I was glad to see some of the Ford Cobras and Mercury Cyclones there, too. Somebody asked, "Why is that?" And I said, "Well, all of us that drove for Chrysler Corporation, we never knew what the back of one of those things looked like!" I was just cutting up — I followed many of them, but the Daytonas and SuperBirds were so strong that it was like they were years ahead of their time.

With the Daytona, we wanted to be the first to take an official time over 200 MPH of any kind of car. We went to Talladega to do the run and from the time we unloaded the car we knew we had something special going. To be the first to run over 200 MPH officially, that is a record that nobody else can ever break. It's important to me, and to do it way back when makes it really special.

NASCAR changed the rules for the 1971 season and the wing cars were gone. I'll tell you the honest truth — I never worry about rules changes. It's kind of like people saying, "It's going to rain today." Well, it's going to rain on everybody. But I hated to see the wing cars go. They had so much potential. They were so good NASCAR got rid of them — does that tell you anything?

Even today, I really have an attachment to the wing cars. They still demand a certain amount of attention whenever you see one. If I pass one that somebody has out on the road or I go to a car show and see one, I point it out to my grandchildren and say, "Gosh, what a great race car that thing was . . ."

Buddy Baker
December, 1994

Introduction

Speed. Power. Beauty.

Summed up in one word: muscle.

The muscle car era of the 1960s and 1970s was a time when the American passenger car underwent a bizarre array of transformations and transmutations. The one common element all muscle cars shared was a hefty powerplant rumbling under the hood. Beyond that, however, cars were stretched, squeezed, lowered, raised, lengthened, shortened, and modified to suit any number of purposes — whether it be for actual performance gains or just for looks.

But there were two muscle cars which rose above all others. Not only did these two cars have a reputation for fierce performance, but their styling could cause a crowd of onlookers to gape in slack-jawed amazement. When one of these cars came down the street, people stopped whatever they were doing and stared. Kids would ride their bikes right off the sidewalk. If the car was parked, you could find it by looking for the crowd that inevitably surrounded it. Or by looking for the wing.

Yes, the wing. These two cars had wings. Not little spoiler wings like those you found riding the rear decks of the Mercury Cyclone Spoiler II — these two cars had towering appendages that rode in the air above even the roofs. People had never seen anything like it before.

In the front, these two cars presented a rounded snout that jutted forward and sliced through the wind — a precursor to the nose found on many modern performance cars.

Today, people might wonder why these two cars needed such odd features. People back then knew.

It was the National Association for Stock Car Auto Racing's (NASCAR) Grand National series that led to the creation of these two cars. Back then, in 1969, everybody believed in the phrase "Win on Sunday, sell on Monday." So did the manufacturers. The manufacturers, especially Ford and Chrysler, liked nothing better than whipping each other on the racetracks where NASCAR competed. And if the win came during one of the biggest races held at the fastest superspeedways — races like the Daytona 500 — well, the bragging could get as intense as the competition on the track.

Ford and Chrysler didn't like to lose and they didn't particularly care for each other's racing programs either. The situation was crystal clear — when one manufacturer won, the other lost. Defeat on the Grand National circuit had an immediate consequence: the all-out search for more speed to turn the tables would urgently leap to the next level.

The search for speed wasn't easy or cheap. It was a bottomless pit that consumed money and man-hours with a promise of quicker lap times just ahead.

Car bodies were tinkered with, more brutal engines were created. And then the guys at Chrysler took the next step. "Let's not concentrate on what we've got now, let's think about what we need to win."

They thought and researched and went to the wind tunnels and the racetracks. They tested and tuned — and they created two models of cars that were like no other car before or since.

The street versions of these cars rolled onto America's highways and instantly became the most astonishing vehicles of a remarkable era — no small achievement when you consider the wide spectrum of exotic and powerful muscle cars available during the era. But it was on the speedways where these two cars had an even greater impact.

The Chrysler teams unleashed them on NASCAR founder Bill France's sanctioned racetracks, and week after week, the late afternoon sunshine cast the shadows of wings across victory lanes.

These two cars were so unusual looking that they began to make "Big Bill" France uncomfortable. They were already making Ford uncomfortable. And then, just when it seemed like Ford was ready to match the stakes of the "expenses be damned" game Chrysler was playing, the manufacturers suddenly quit the racing game altogether. They packed up their engineering teams and slammed shut the factory checkbooks and retreated to Michigan, leaving the race teams to fend for themselves.

But the manufacturers did leave behind a legacy. They left behind memories of a time when racing was an all-out war between giant automotive corporations, memories of a time when races were as likely to be won in wind tunnels and engineering meetings as they were on a racetrack, and memories of two very special race cars.

These two cars can now be found in museums and in the conversations of the men who built them and the men who raced in and against them — men who will never forget how great an achievement these cars really were.

And now, when their owners carefully guide them down the road, the street versions of these two cars rub shoulders not with muscle, but with minivans. Still, no matter how much has changed in America since the creation of these supercars in 1969 and 1970, one thing is certain — these two cars are guaranteed to turn the head of every person they rumble by.

These two cars are the Dodge Charger Daytona and the Plymouth SuperBird.

Chrysler collector Bill Stech's immaculate supercars—a Daytona and a SuperBird.

Muscle Mania and the Need for Speed

In America in the late 1960s, anything seemed possible: man was headed for the moon, Jimi Hendrix was re-inventing how the electric guitar should be played, college campuses were in upheaval, Joe Namath led the New York Jets to a stunning Super Bowl victory over the Baltimore Colts, Americans were fighting in a vicious jungle war on the other side of the globe, fashions had taken a turn for the bizarre, and in the automotive industry, when it came to engines there was a simple rule—bigger was better.

The era of the muscle car began by capitalizing on the American public's fascination with automotive power. The first indications of the coming of true muscle were seen in the mid-1950s when car manufacturers began to offer hefty powerplant options in their full-size models.

The 1955 Chrysler 300 was the first of the Chrylser cars to really flex its muscle. The 300 was powered by a 331-cubic-inch motor that made it the hottest thing going on the American road. The car was developed from a proposal by engineer Bob Rodger, and while earlier models from Chrysler had offered the occasional high-power option, the Chrysler 300 was the first model to make no bones about its capabilities and intentions.

Of course, success breeds imitation and soon all of the manufacturers were developing high-powered vehicles of their own. With the introduction of the Ford Mustang in April 1964, "pony cars" became the rage and smaller and sportier vehicles made their mark on the muscle car scene.

But there was a drawback to the pony cars — there often wasn't enough room in the engine compartment to hold a fuel-guzzling, big-block, large-displacement powerplant. Real muscle required real room, so the vast majority of muscle cars were based on mid-size models.

In 1966 Dodge introduced what would become one of the greatest of all muscle cars, the Dodge Charger. The Dodge Coronet had been one of the division's biggest sellers, but sporty styling and fastbacks had become popular in the industry and the Coronet lacked both. Dodge designer Bill Brownlie proposed adding a fastback to the Coronet, and the change was grafted on a Coronet body and dubbed the Charger II for the auto show circuit. Public reaction was so positive that the Dodge Charger made its showroom debut shortly thereafter, retaining much of the Charger II's styling flair.

In its first model year, 468 of the Chargers received the ultimate in muscle — the 426-cubic-inch hemi-head engine, capable of putting out more than 400 horsepower. The Hemi was already legendary, but added to the sleek Charger body, it created a potent new combination ready to bring thunder to the streets.

As was standard at Chrysler during this time period, models generally received a full-scale styling revision every two to three years, and the Charger

Opposite: This ⅜-scale model of the first generation Dodge Charger signaled the beginning of a new era of stock car racing aerodynamics, during which wind tunnel testing became as important as racetrack testing.

was no exception. The 1968 model was a beautiful creation, with lines that gave the illusion of high performance even when the car was parked.

Over in Chrysler's other division, Plymouth had developed muscle of its own. The idea was to create a no-frills muscle car, one that might even be considered an economy muscle car. But one aspect of this car certainly wouldn't be bare bones — there would be no skimping in the engine department. Much as Dodge's Coronet had spawned the Charger, the Plymouth Belvedere was the parent car of this muscular creation. In a stroke of marketing genius, product planner Jack Smith named the car after a cartoon — the Road Runner.

When it made its debut in 1968, the Plymouth Road Runner went so far as to duplicate the cartoon character's distinctive "beep! beep!" voicing for the production car's horn. But the horn wasn't the only thing under the Road Runner's hood — like the Charger before it, the Road Runner could be equipped with the Hemi. Almost instantly, the Plymouth Road Runner became one of the most popular muscle cars on the streets.

But the street action was just one variable in the muscle equation. The Charger and Road Runner — technically referred to as B-body cars — were the front line warriors of the Chrysler Corporation's NASCAR Grand National efforts. On the superspeedways and short tracks of America, the automobile manufacturers were battling for supremacy — and bragging rights.

"Win on Sunday, sell on Monday" — that was the philosophy of factory-backed racing. The spectators in the stands built fierce loyalties to drivers and makes of cars, but most probably didn't realize what went on behind the scenes. What you saw at the racetrack was just the tip of the NASCAR iceberg. What you didn't see was the factory engineering armies toiling away to put their brand on top. Much of the work was done far from the racetracks, but was absolutely crucial to NASCAR success. A win in the Daytona 500 was strategic enough and valuable enough from a marketing standpoint to justify the countless dollars that were being poured into the racing efforts.

In the 1960s, factories were fully supporting and maintaining their race teams. "They had specific places where they had engines built and cars built and they had people running them so if you needed an engine or you needed gears you'd just go pick them out," recalls Grand National driver Dick Brooks.

"There was a lot of information shared," notes NASCAR champion Ford driver Cale Yarborough. "Ford had engineers to work with every Ford team so a lot of information was shared to make each of us as good as we could possibly be. And we had a lot of success with it."

"If you didn't have a Ford or Chrysler factory ride you were wasting your time," Daytona 500 winner Pete Hamilton says. "You had to have a factory ride."

"They were like race teams themselves that just didn't have a race car," recalls crew chief Harry Hyde of the help Chrysler provided to his race team. "They were a team support for the teams, and they were the best bunch of guys that you would ever want to meet. Those sons-of-guns came out to the racetrack and they weren't all dressed up — they had cobble-toed shoes on and them dudes just plowed into it. They helped us in many, many, many ways that we don't see today. We were really close to them and they were close to us. They wanted to service you, to see what they could do for you."

The factories had outposts which they used to distribute their racing equipment. The center of Ford's Grand National operations was the Charlotte, North Carolina-based Holman-Moody shop. Chrysler's distribution location was Nichels Engineering in Highland, Indiana.

"At that time there was Holman-Moody and there was Nichels Engineering," explains Dick Hutcherson, who was general manager of the Holman-Moody operation and supervised David Pearson's championship teams in 1968 and 1969. "If you wanted to run a Ford product you went to Holman-Moody. Chrysler products, they were built at Nichels. Even most of Richard Petty's cars were

The 1968 Charger (left) was a totally new design that Dodge hoped would out-perform the early Chargers like the one driven in NASCAR Grand National competition by Sam McQuagg (right).

built at Nichels, and then Petty might re-work them some. Now there's any number of places you can go get race cars — I can think of five right off the top of my hat. It wasn't like that back then. Ford had their design, and Chrysler had their design."

Before the parts or cars ever arrived at Nichels Engineering, however, they had passed through the hierarchy of the Chrysler Corporation's racing operations. This program was administered through the Product Planning department, and Ronnie Householder was the manager who oversaw the operation. But the results seen on the speedways came from the labor of the Special Vehicles Group, formed in 1964, and from the Chrysler aerodynamics people.

Brute force from huge engines was the traditional way to make speed, but aerodynamics was beginning to be seen as a new way to get the same results. The idea of seriously studying aerodynamics in automobile racing was a fairly new one, and many of the Chrysler people had moved into the automotive world from the aerospace world.

"I'd come out from the missile division — my title there was aerodynamicist," recalls Chrysler's John Pointer. "I was working on hypersonic flow around nose cones. The missile went toes up and I was laid off for about three or four months. I saw an ad in the paper — Chrysler wants aerodynamicists, automotive. I went out to the proving grounds and they told me, 'What we want you to do is build or develop an experimental technique measuring the drag of the car.' I knew nothing about aerodynamics below Mach 3 and then only around anti-symmetric bodies at zero angle of attack for re-entry. I thought, 'Well, I'll just kind of do a literature search and see what's been done in the past.' I spent about three weeks and found nobody knew how to do this."

"It was very exciting," says Bob Marcell, another Chrysler aerodynamicist. "I came out of college, and I worked in aerospace for a couple of years in research at University of Michigan. I got to work there on space program stuff, re-entry vehicles. But the stuff we did there at Chrysler was damn exciting because we definitely were doing some things that there hadn't been a lot of work done on before."

Special Vehicle Group's Larry Rathgeb, who was in charge of engineering in the stock car racing program, notes, "Chrysler itself did a tremendous amount of ground-breaking work with aerodynamics very early in the game, back in 1963 and 1964. I can remember looking at aerodynamics in 1966 with the new Charger with the big sloped back on it."

"That was a real challenge," remarks Bob Marcell. "We started out with some pretty subtle things. I remember coming into the group and I was a junior aerodynamicist. We had an aerodynamics group of about five. Some of us worked on regular passenger cars but I was one of those who worked pretty much on the race program. I remember going with Larry Rathgeb, and we were just doing little things like rounding off radius on the hood. We put little lip spoilers at the end of the deck, and we started playing around with backlights. We started flushing up glass and looking for small improvements. And it was kind of comical, because it really escalated!"

Although it is not unusual for sites like the modern Chrysler Technical Center in Michigan to have a wind tunnel, in the 1960s the Chrysler aerodynamics team had to travel off-site for tunnel work. Two primary locations were used—the Wichita State University 7' x 10' tunnel, and the 16' x 23' facility at Lockheed-Georgia.

For studies done at Wichita State, ⅜-scale models had to be used. Why that size? "We used ⅜ because wind tunnels like Wichita State have 5' x 7' type cross sections," answers Bob Marcell. "One of the keys there is the larger you make the model relative to the tunnel the more corrections you have to make and the more uncertainty there is in your data. So that was about the right match to minimize the buoyancy and blockage and other aerodynamic corrections. Back then there wasn't an awful lot known about that. Gary Romberg and I spent virtually months researching what the aircraft people did, but because we had ground effects it was a lot tougher."

The Lockheed tunnel, constructed during the months leading up to NASA's Apollo missions, had never been used for automobile tests.

Dick Lajoie was also working in aerospace; he was at the Chrysler Corporation Defense and Space Division in Huntsville, Alabama. Product planner Dale Reeker called Lajoie and asked him to look at the Lockheed facility that was being commissioned.

"So John Vaughn and I flew out to Atlanta and looked at the tunnel," remembers Lajoie. "We talked to somebody and they said, 'Right now, all we've got is this big hole in the ceiling where they drop the aircraft models.' We said, 'Can you do that with cars?'"

Lajoie and Vaughn went to Michigan and showed everyone photographs of the tunnel. Chrysler wanted to take advantage of the Lockheed tunnel's full-size testing capabilities, but there was a big problem. The hole in the tunnel through which the test vehicles were to be lowered was 14 feet long — and the cars measured nearly 18 feet. The problem had just one solution — a cradle would have to be tooled to angle the car and get it through the hole and into the tunnel.

Bob Marcell was involved in the first tests at Lockheed — a study of drafting. When two stock cars run close together on a superspeedway, they can slice through the air faster than one car alone. Knowing what the air will do around your car at high speeds can be a big advantage in stock car racing. Engineering's H. Paul Bruns had allocated $50,000 for the testing.

Wind tunnel testing with ⅜-scale models at Wichita State University.

"So we were going to do drafting tests," says Marcell. "Well, they charged $500 per hour down there whether you had the wind on or not. Hell, we spent two days just putting in the ground plane for the cars, trying to get the balance system to interface correctly. And I remember calling Paul and saying, 'Paul, we're here to find out about how cars draft, but I spent all of the money and we haven't even turned on the fan yet! I need another $50,000.' Back in the late '60s that was a lot. But we got the money and we ran there for two or three days. It took us forever to get our data reduced because they didn't have some of the capabilities that Wichita had because they hadn't run cars before."

The urgency of these tests could be explained simply — the 1968 Dodge Charger was being defeated by Fords in the NASCAR Grand National series. While Richard Petty was still successful in his Plymouth cars — coming off an amazing 27 wins in 1967 and on his way to 16 for 1968 — the Charger performance was disappointing, even though things had looked just fine before the season began.

"I had one Sunday with one prototype street car to evaluate that Charger, and it looked better than anything we'd done," recalls test engineer John Pointer. "We'd done some computer modeling and we figured this car ought to be able to lap the track at 184 MPH."

This would have put the Dodge teams in good shape since the pole qualifying speed at the 1967 Daytona 500 had been Curtis Turner's lap of 180.831 MPH.

"One hundred eighty-four MPH was faster than the track record so everybody sat back and rubbed their bellies — we were in good shape," Pointer continues. "The only thing was the drivers reported that at speeds over 100 MPH the front end felt funny. I was not there at the first test but I was told Buddy Baker was blasting down the front straightaway, really having a good time, and he cuts the wheel going into the corner, and nothing happened. And he cut it further and nothing happened. He cut it back the other way and nothing happened. Finally he decided to get out of the throttle. He took a big spot out of the chrome on the right tip of the bumper — ticked the wall. We were in trouble. Petty in his '68 Plymouth was the only car that could stay with the Fords."

The problem was twofold. At the front end, the 1968 Charger had a recessed grille that acted like an air scoop. Air would rush into this area and collide with the vertical surface. The air had nowhere to go, causing havoc with the car's handling on superspeedways. There was nearly 1,250 pounds of front lift at 180 MPH.

Equally troublesome was the roof line. Although in profile the eye-pleasing Charger appeared to be almost a fastback, the rear window was nearly vertical. Between the swept back roof pillars and the vertical glass, a tunnel effect around this backlight area created dangerous lift at high speeds.

The use of a front spoiler reduced the front lift to 500-600 pounds at 180 MPH. Though this improved the handling, it was far from a perfect solution because it created additional drag.

"The idea was to get a reduced drag," explains George Wallace, a key member of the Special Vehicle Group. "We knew that the nose and the backlight were by no means optimum for the vehicle, but it was what we had. The Fords were showing us they definitely had either done their homework or they had stumbled onto it — I think it might have been a combination of the two. But the fastback Torinos and the similar Mercurys were definitely faster than we were. We had more power in the Hemi, but it wasn't enough."

Chrysler rushed to find a solution, and the result was the 1969 Dodge Charger 500. Under NASCAR rules, there was a production minimum of 500 units to qualify a street vehicle for competition — hence the Charger 500's name, although most sources agree only 392 ever made it onto the road.

"The Charger 500 was shown to the press in June of 1968 at a press preview out at the proving grounds," recalls John Pointer. "It was actually a '68

Yarn tests in the wind tunnel reveal air flow over the cars. Note how yarn on the 1968 Charger rear window area (top) rises straight up, while the Charger 500's fastback substantially lessens the rear end lift problem (bottom).

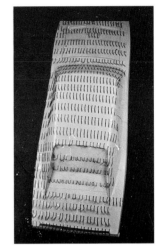

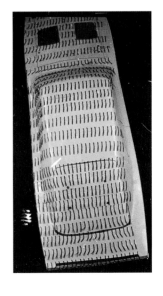

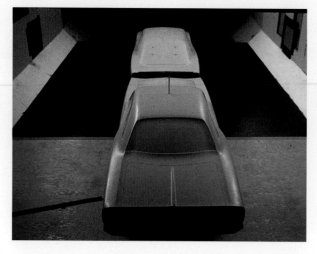

Wind tunnel testing for the Dodge Charger 500 program.

Charger street machine that was done over to be a Charger 500. If you look real closely at it you can see it had the '68 model year side marker lights, not the '69's."

The Charger 500 was based on a sketch drawn by John Pointer during a Daytona testing session. Pointer's design aimed to solve the aerodynamic problems that plagued the 1968 Charger. The Chrysler team borrowed the grille from the 1968 Coronet — the Charger parent car — and mounted it flush with the Charger's front end, effectively eliminating one problem. The solution for the backlight area was to make the car a modified fastback with a flush rear window by means of a plug to replace the buttress-back design that had created so much lift.

The street model Charger 500s were transformed from standard models by Creative Industries, a Detroit firm that had taken on special projects for Chrysler in the past. But Dodge was mostly concerned with racetrack performance, and they were eagerly looking forward to the 1969 Daytona 500.

While Chrysler was revising the Charger, Ford was not standing still. Their sleek Torino was being made even sleeker through the work of Ralph Moody, visionary partner of the Holman-Moody Ford racing operations. Moody had cooked up a longer and lower nose for the Torino, and utilized a flush grille as well. Ford's racing boss, Jacques Passino, was pleased with Moody's handiwork, and the car was christened the Talladega after the new Alabama International Motor Speedway.

"Ralph Moody, he's the one who came up with the Talladega," remembers Dick Hutcherson. "Ralph was a pretty sharp individual and it was designed right there at Holman-Moody. Then it was taken to the wind tunnel, of course, but it was really Ralph's design."

A showdown loomed at Daytona International Speedway, where the Dodge Charger 500 and Ford Torino Talladega — along with its nearly identical Mercury cousin, the Cyclone Spoiler II — would battle to win the most important stock car race of the year, the 1969 Daytona 500.

Tension was high as the race neared. "There was always worry and concern when they brought something new out," Holman-Moody's Hutcherson says of Chrysler. "I would say that maybe engineering-wise Chrysler was a little ahead of Ford, but we pretty well held our own with them."

That they did. In the first of two 125-mile qualifying races for the Daytona 500, first, second, and third went to a Ford, a Mercury, and a Ford. David Pearson had won the race, followed closely by Cale Yarborough and Donnie Allison. In the second qualifier, the Dodges established themselves with a one-two-three sweep by Bobby Isaac, Charlie Glotzbach, and Paul Goldsmith.

When the Daytona 500 began, the competition was as close as had been expected. The Chryslers and Fords did indeed stage a fierce battle, and it all came down to the last lap. LeeRoy Yarbrough, whose crew had mounted a softer, more adhesive tire on his Talladega, fought past Charlie Glotzbach's Dodge on the last lap. "Chargin' Charlie" made a desperate attempt at a slingshot pass off the 31-degree banking of turn four, but his Charger 500 came up just short. Yarbrough's Torino Talladega had won the Daytona 500 by less than a car length.

"The Ford people had created the Talladega car, which was superior," Special Vehicle Group's Larry Rathgeb admits. "It was a smaller car, and it was superior in aerodynamics and in performance to the Charger 500."

"Overall we weren't any better off than we had been the year before," remembers George Wallace. "Chrysler management really wanted to win, and they particularly wanted to win at Daytona. Just as in Indy cars — winning at Indy is everything. If you won Daytona, the rest of it was of minor importance. Everybody knew who won at Daytona. The championship and all that was nice, but winning at Daytona was the name of the game," Wallace emphasizes.

Despite the close finish in the Daytona 500, the loss was considered a disaster and faith in the Charger 500's capabilities plummeted. But work had already begun on what would become one of the most amazing cars ever to race on a Grand National speedway — the 1970 Dodge Charger race car. At the highest levels of Chrysler racing, a decision was made.

Chrysler's John Pointer clearly remembers the result of that decision. "The next generation Charger stock car was intended to be a 1970 model, and the follow-up we needed — suddenly it became a '69 model."

What was the reasoning behind putting the entire Chrysler racing operation into the accelerated pace of a crash program?

"Because the '69 lost . . ." says Pointer.

"And the bottom line is win," Dick Lajoie finishes.

Don White next to his long and low Dodge Charger 500. This new variation was good — but not good enough.

How Chargers Grow Wings

"Chrysler was good at coming up with something exotic — they'd always been through the years," Grand National driver Richard Brickhouse says. "They were so determined to beat Ford, they dreamed this car up on their own. This was their baby, and they knew what it would do. A lot of people went to fussing and complaining that these cars got too exotic, too far out, didn't look like race cars, blah blah blah — but they knew what they were doing. They meant to win some races."

"Chrysler, they were very competitive in racing," agrees Charlie Glotzbach. "They wanted to be number one and they didn't care what it took to do it."

Chrysler aerodynamicist Dick Lajoie remembers his first meeting with vehicle test engineer John Pointer. "John was working with this mass of metal that had some strange looking wing and a strange front end," Lajoie recalls. "That was the preliminary work on the Dodge Charger Daytona."

Pointer's "mass of metal" was expected to be in shape to win the first NASCAR Grand National event at Alabama International Motor Speedway in September 1969, Chrysler's target date for the Dodge Charger Daytona to be ready to race. The idea for a new race vehicle was born at a meeting between two members of the Chrysler Special Vehicles Group — although at the time of the meeting the Dodge Charger Daytona was still planned as a 1970 model. Larry Rathgeb was the supervisor in charge of circle track racing development, and George Wallace was a technical advisor in both the NASCAR and drag racing programs.

"We sat together and we decided we had to do something to improve our performance," recalls Rathgeb of the late 1968 meeting. "We called and we talked to two people — John Pointer at the proving grounds, and Bob Marcell of aerodynamics. I asked them if they couldn't look over what we had in the way of 1969 vehicles to see what vehicle could

Opposite: Working with full-size cars was a lot harder than working with scale models. This photo shows the special cradle that was used to get cars into and out of the Lockheed-Georgia wind tunnel.

Left: Illustrations from a Society of Automotive Engineers (SAE) paper entitled "The Aerodynamic Development of the Charger Daytona for Stock Car Competition."

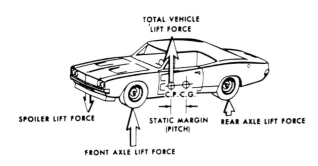

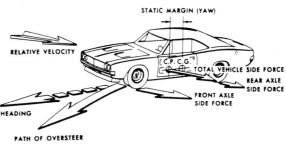

Larry Rathgeb asked John Pointer and Bob Marcell to come up with some sketches to improve the Charger design. This was Pointer's radical concept, and Marcell was thinking along the same lines.

best be altered in some manner to improve the aerodynamics enough to be better than the Ford. I asked them if they'd make up some sketches or drawings and then meet with us.

"Then we called the Plymouth and the Dodge people — that was Dale Reeker with Dodge and Jim Stickford with Plymouth. We asked them if they would attend this meeting, that we wanted to talk to them about creating vehicles that would be superior to the Fords," Rathgeb continues. "When the meeting occurred, everybody was there except for Stickford. So we called him and said, 'Jim, did you forget about this meeting?' And he said, 'No, I talked to my boss and he said that we have Richard Petty. We really don't need engineering's help since we have Richard Petty for the Plymouth.'"

Attending this pivotal meeting were Special Vehicles Group's Larry Rathgeb and George Wallace, Dodge product planner Dale Reeker, and aerodynamics experts John Pointer and Bob Marcell. The first order of business was to look over the drawings of possible enhancements envisioned by Pointer and Marcell.

"Marcell and I had both done some sketches," remembers Pointer. "We'd both come up with basically a smoothing of the front end to cut the drag and reduce the front lift so we didn't need a big front spoiler to balance the car."

"The drawings were very similar," says Marcell. "Both of us were looking to do something very dramatic with the front end. We also had to do something with the backlight."

"The two guys came up with drawings that were very similar and yet they did it on their own — they had not gotten together prior to this or talked," points out Rathgeb. "Each of the cars had a sloped nose and a wing — one had a double wing in the back and one had a single wing. But the thing that was interesting was that they'd both come up with similar concepts for a better aerodynamic car from the same vehicle — the Charger 500."

Rathgeb and Reeker were excited about the potential for a winner, and began the process of gaining the necessary approval to begin a full-scale development program. The first stop was the office of the chief engineer of Chrysler Product Planning, Bob Rodger, the man who had proposed the first Chrysler muscle car, the Chrysler 300.

"We said that we'd like to go ahead and create this thing," Rathgeb remembers. "Rodger said, 'Well, I'm glad somebody is still interested in trying to better our products for racing. Why don't you go ahead and see what you can do about it?'"

Proceeding up the corporate ladder, the next stop was to be the big payoff — vice president and general manager of Dodge, Bob McCurry. Sketches in hand, Reeker and Rathgeb met with McCurry.

"He looked at the thing and he said, 'God, it looks awful,'" Rathgeb recalls. "Then he said, 'Will it win races?' And I said, 'Yes, it will win races.' And he said, 'Well damn it — go ahead and build it!' And he said, 'If anybody gets in your way just let me know and I'll keep the path clear for you.' So we went ahead and did it — we built the car."

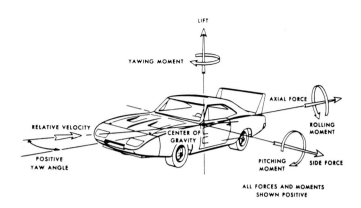
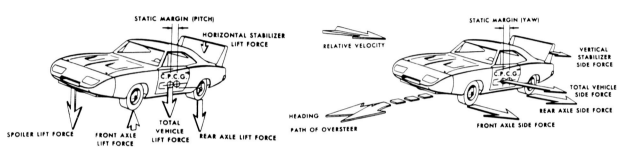

Illustrations from an SAE paper entitled "The Aerodynamic Development of the Charger Daytona for Stock Car Competition."

The Dodge Charger Daytona program did not follow the rules used by major automobile manufacturers to develop new models. A proclamation stating that the Daytona had to be ready to race by September 1969 was issued in mid-March. What would ordinarily have been a massive procedure was instead a lean, three-pronged program. One effort directed by John Pointer would be underway at the Chrysler Proving Grounds. At the same time, the Aerodynamics Group — which included Bob Marcell, Gary Romberg, Dick Lajoie, Frank Chianese, and John Vaughn — would seek solutions in the wind tunnels. Finally, product planner Dale Reeker was to coordinate Creative Industries' efforts to build the required number of street models to qualify the car for NASCAR competition.

The goal of the Daytona program was to come up with a 5 MPH increase in lap speed on the superspeedways. For an engine to generate such an increase would require 85 additional horsepower; the Chrysler team set its sights on achieving the same results by reducing aerodynamic drag by 15 percent.

"Most of the sketch and design work went on a crash and burn basis to get the car ready for 1969," recalls aerodynamicist Gary Romberg. "We started late to be able to do a race car. One of the reasons that the whole program was successful was because it was as small as it was. It was a little team that crashed and burned to be able to get that Dodge done."

"The Daytona had to be introduced by the 15th of April of 1969," says John Pointer. "There had to be 500 cars in hand to customers by the first of September. And we had eight and one half months to do that."

"The bottom line is that we had wind tunnel tests going on concurrently with the road tests concurrently with the fabrication of the car," notes aerodynamicist Dick Lajoie. "Everybody was on the telephone, giving results, so that John Pointer

could check it out on the track to see if it proved out the experimental data."

Pointer's initial test bed at Chrysler's Chelsea, Michigan, proving grounds was a car with a checkered past as far as NASCAR was concerned. It had been driven by Charlie Glotzbach in the 1968 Firecracker 400.

"For the July 1968 Daytona race we built up what we called the two-by-two cars," explains George Wallace. "The body was lowered on the chassis by two inches at the front and at the rear. Aerodynamically there were no changes other than the fact that the car was lower. Things went moderately well but they weren't world-beaters, and then NASCAR banned them. We had two Dodges and one Plymouth, and one of the Dodges ended up being the mule that was used by John Pointer at the proving grounds to develop the Daytona."

After having its race engine replaced by a production Hemi to allow the use of standard gasoline, the proving grounds "mule" — auto industry slang for a car already in production that is updated to a future model shape — underwent intense alterations.

As Pointer explains, "The NASCAR rules are just one of the variables you have to contend with, but physically speaking there is no way to predict what is going to happen. You cut and you try it. You put yarn on the car or smoke streamers in the tunnel and make some guesses and start working with it. The nose work on the Daytona started in January of 1969 with scrap aluminum and angle iron."

With highly detailed ⅜-scale models at the Wichita State wind tunnel and full-size data from the Lockheed-Georgia facility, the aerodynamicists had evaluated two options for the nose which Pointer began working with. Eventually a sloping nose that stretched 18 inches from the bumper supports was chosen over a nine-inch variation. The design chosen afforded lower drag, better directional stability, and — for the street model — created a location for headlights.

"I'd go down to Wichita and typically take a mechanic and a clay modeler and then we'd use a couple of people there," Bob Marcell says of working on the models. "One thing we could do quicker than John Pointer was really refine the design. We'd be getting templates back to him, and he'd be outfitting them on the cars to see how they did. We had data reduction fairly quickly. I think it worked out fairly well — there was never much disagreement. We honed in on the design and John kind of followed suit on that and we'd end up pretty close on where we estimated we'd be. We could get within a couple of MPH of where the car would actually run — and when you're up in the 190s to 200, that's pretty accurate. It was a pretty small team, but we did a hell of a lot of work."

The most noticeable feature of the Daytona — its wing — symbolizes the cooperative efforts between

 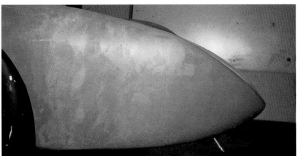

Opposite and left: While full-size testing was underway at the Lockheed wind tunnel, studies of unusual new nose configurations modeled in clay were taking place at Wichita State.

the wind tunnel operations and the proving grounds work. Initial studies indicated that the best place to mount the wing was directly over the rear axle, but practical considerations dictated moving the device aft.

Pointer had been investigating the best height to mount the aerodynamic aid. "I did a parametric study on the rear wing on the Daytona. We had killed the front lift, but drivers feel more comfortable if there is an imbalance between the front and rear normal forces. So the spoiler would have had to have been the size of a barn door to balance the car again," Pointer says. "That would have put us back behind where we were with the Charger 500. A wing was the way to go, so I built an arbitrary wing six inches off the deck, 12 inches off the deck . . . and somewhere around 15 or 17 inches off the deck you stopped getting any benefit."

There was just one problem — with a wing at that height there would be no way to open the trunk on the street model Daytonas. "It had to be 23 inches off the deck so we could get the trunk lid open far enough to get the spare tire out of the bottom of the trunk," continues Pointer. "At that point I was just working with posts that held the wing up. Then we said, 'Why don't we make this a streamlined fairing, especially since it's got to be that high — maybe it'll help the stability?' I had no way of directly measuring that at the track. In the tunnel, they could make more radical changes quickly and they could yaw the model — but I couldn't. So it was back and forth on the phone, 'Hey, what did you find?'"

While a wing had been part of the plans for the Daytona all along, figuring out the best way to use it had fallen to the men toiling in the wind tunnel.

"Gary Romberg and the aerodynamics people came up with the idea of using a Clark Y airfoil upside down, which pulls the car down instead of lifting it," says Larry Rathgeb.

According to Rathgeb, the upside-down airfoil and the streamlined vertical supports were a tremendous help. "By doing all that we did, we moved the center of pressure somewhere back just forward of the center of gravity and you really had to steer the car," Rathgeb notes. "It was a very, very stable car because the side forces never tried to move the car, either move the nose out or move the tail out."

"If the tail did try to come out, as soon as it began to come out it would expose more and more of the wing uprights to the air. As the car begins to move, to arc about its center of gravity, the air sees more and more of the opposite upright. As the car moves further and further that piece gets bigger and bigger and so it just holds it in. It's perfect — the whole thing was ideal."

The project's accelerated pace was highly unusual for the automotive industry — as was the fact that the styling department was not getting its say. "We don't build cars like that," Gary Romberg points out. "This car was funded through marketing, and the real goal in this thing was win or die. There was only a small number of people, mostly in engineering, that were a part of this. Styling was on the outside of this — that wasn't normal. Styling at the time was very powerful."

But even more powerful was Bob McCurry, the Dodge vice president and general manager who had given the Daytona project the go-ahead while promising to keep the path clear for rapid development.

As John Pointer recalls, "Dale Reeker, Dick Lajoie, and I were at Creative Industries and there was a stylist who wanted to help with the Daytona. He wanted to take out his boning knife and start carving clay on this front end. We said, 'No, no, no,' and he did a lot of muttering. This was back when styling was in Building 128, and a few days later we were there and this stylist started calling Dale every name in the book because we wouldn't let him touch the Daytona. McCurry came up while this guy was going on with his tirade and said, 'What's going on?'

"And so this guy launches into it again," Pointer continues. "I mean, I spent two years in the U.S. Army and I never heard a master sergeant come out with a string of stuff like this guy did — he went on and on and on. He finally paused for a breath. And McCurry looked him right in the eye and he said, 'I don't give a shit what it looks like. It goes fast. If you can't help, get out of the way.'"

"There was some bad feeling inside due to styling's not being part of this thing," says Romberg. "There were a lot of real bad predictions —'That car's not going to sell,' 'It's an abortion,' 'You're going to wreck the image of the company' — and all those kinds of things were said. And then when the car came out we had more orders for them than we could even handle."

Announcing the Daytona's existence and taking dealer orders was one thing — keeping the car a secret until it was strategically safe to reveal it was another.

"We knew that once we started, word would get out," admits George Wallace. "There was then, and I suspect there still is, a pretty big underground in Detroit of finding out what competitors are doing. Creative Industries was actually doing the work of the modifications, and Creative did work for Chrysler, Ford, and GM. Half of the people from a company who would come around to check on how their projects were coming were really looking to see what the competition was doing . . . When you're doing that big of a project you really can't keep a blanket over all of it."

Sometimes unwanted attention was unavoidable. When Bob Marcell went to test at the Lockheed wind tunnel in Georgia, he took full-scale race cars with him. "You get down there in that part of the country, where it's a real hotbed for racing, and we had all of this security. But then when they backed the cars out of the moving van you could hear them

The number 6 Charger driven by Charlie Glotzbach (left) was banned by NASCAR for being too low, so John Pointer inherited it to test Charger Daytona ideas. An artist's sketch (right) captures the embryonic Daytona design.

Chrysler's John Pointer labored to make sure the Daytona's fender scoops did nothing more than produce tire clearance, despite rumors to the contrary.

about 25 miles away!" exclaims Marcell. "If you followed racing, you knew it was a Hemi — so they kind of knew who was in the tunnel."

"With both the production stuff and the racing stuff, the feeling is that if you've kept it secret until you actually start making cars, you've got enough leeway that the competition across town may know you're doing it but they can't have an answer on the street when you've got yours on the street," points out Wallace. "They would be a year away from copying it."

The veil of secrecy regarding the Daytona's existence was lifted on April 13, 1969, when the Dodge Charger Daytona was revealed to the automotive press. The writers were treated to the sight of a Daytona that wasn't really a Daytona — it was actually a standard model Charger 500 with a rush-job fiber glass nose and spoiler. The press was also shown Pointer's mule, a winged stock car mock-up which still had posts elevating the airfoil as opposed to the streamlined fairings which would eventually be developed.

Curiously, one aspect of the press preview seemed to cause endless fascination and speculation among the witnesses. Why did the wing car have rearward facing scoops on the fenders above the front wheels? To this day, articles and books about that era still come up with a bizarre array of explanations for the scoops, ranging from alleged aerodynamic benefits to engine heat ventilation. In reality, the purpose of the scoops is considerably less exotic.

At the NASCAR superspeedways, front tire clearance in the wheel well was often a problem. The tires would be forced into contact with the fenders when the car passed through a high-banked turn at nearly 200 MPH. To solve that problem on the Daytona, Larry Rathgeb came up with an idea — a backwards facing air scoop that would do nothing more than provide room to prevent the tire from coming in contact with the fenders.

"I worked furiously on those things to make sure they didn't do anything aerodynamically," John Pointer says. "They were just there to provide an extra inch or two of tire clearance."

As Gary Romberg notes, "There were myths around those things all the way to the racetrack for years! All during the time that the cars were on the racetracks, that was supposed to be one of the major tricks. And that was no more than tire clearance!"

Now that the car had been revealed to the public, the final big push was underway to ensure that the Daytona would make its September deadline.

"What we had was an organization that placed people through each of the departments, laboratories, design rooms, and whatever — wherever we thought we needed help we'd place a person," explains Rathgeb. "Whenever we needed help we could go to those organizations and get that help with no problem. We had the whole corporation behind us — everybody! Whatever we needed, we had. We had an axle breakage problem. We took the axle to the gear people and said, 'Hey, we've got a problem here. What can we do about it?' Well, in three days they had the problem solved. 'You cut back the splines here, reduce the diameter in this area, increase the diameter here, change the hardness in this area, and it's done!'"

As another great mission — the Apollo 11 lunar landing — was coming to fruition, the real testing of the Dodge Charger Daytona began. Until this point, Pointer's converted two-by-two Daytona mule had been restricted to speeds below 120 MPH as different aspects of the car were tested. The results were calibrated to equate with the higher speeds of the Grand National tracks. Now the mule was put out to pasture as a new, winged engineering race car took the place of the old modified Dodge for the first high-speed tests of the Daytona.

Chrysler Grand National stars Charlie Glotzbach and Buddy Baker were selected as the test drivers. Each had won in 1968 and had numerous top-ten finishes in the old Chargers.

"I think I could relate things back to Rathgeb," Glotzbach says of his selection. "I've always been pretty mechanically inclined. I used to build my own cars and work on them when I was running the short tracks, and I could tell him what the car was doing. A lot of the drivers couldn't. I think I ran every test that was run on Daytonas as far as oval tracks."

Why was Baker the other choice? "I've heard it said, 'Number two hat, number fourteen shoe!'" Baker jokes. "I guess it's because I raced, I qualified, I practiced the same way — wide open. When they wanted to evaluate things I feel that Charlie and I would go out and run — if it was a 500-mile test, we'd run 500 miles as hard as the car would go."

The bottom line reason for the selection of Glotzbach and Baker, according to Larry Rathgeb, was simple: "They weren't afraid to go fast!"

The day before the testing was to begin, John Pointer took Baker and Glotzbach out to the Chrysler Proving Grounds at Chelsea, Michigan. "I took them out on a quick tour of the joint, and then they were making sure everything fit properly," recalls Pointer. "Now Charlie Glotzbach is about average size, and Buddy Baker is enormous. Back when I weighed 175 pounds he used to pick me up and say, 'How you doin', little buddy?' Anyway, Buddy was sitting there trying to get this helmet on, sitting on a pile of tires and struggling with it. Glotzbach comes up behind him, balled up both his fists like this, and BAM! Both Charlie's feet were off the ground — the full weight of his body was on top of that helmet! He drove it down, Buddy's ears were sticking forward, tears pouring down his cheeks, and he got up and looked at Charlie and made some remarks about Charlie's relationship with his mother, what his mother did for a living, and so on. But Charlie looked at him and said, 'Just trying to help you out, Buddy. I seen you was having a little trouble!'"

The day after this light-hearted initiation, the two drivers would be piloting an untested race car around the five-mile Chelsea track. "You don't have any guardrails or anything — if something was to happen you'd still be flying, I guess," Glotzbach remembers of the huge course. "The turns are made kind of like a spoon. You can go down in the corner and turn the steering wheel loose at 180 MPH and the car would automatically turn."

At the test track, the banking in the turns increased in steepness with each lane. "You'd just move up a lane when you wanted to run faster," Baker explains. "If you were running a little faster than your car felt good you'd just move up a lane, and pretty soon you were running fast enough where you could almost look straight down. It was parabolic, and it was plenty wide." The drivers had six lanes and a maximum banking of 31 degrees.

July 20, 1969 — man is on the moon, and the Dodge Charger Daytona makes its first high-speed runs. On the moon, Apollo 11's mission was progressing smoothly. At Chelsea, things weren't going as well — or were they?

After a detailed inspection of the track surface, tests comparing the Daytona with a Ray Nichels Charger 500 brought in from Indiana began. The wing car was faster than its relative, but just barely — the top speed was a disappointing 194 MPH.

Although most people interviewed for this book believe that speeds at that first test were lower than expected due to a damaged Hemi engine in the Daytona, both Larry Rathgeb and George Wallace say that the slower speeds were by design.

"We knew that we could not keep this test a secret," Wallace reveals. "Too many people had to be involved and somebody would say, 'How fast have you gone?' So we built up a Hemi race engine with a very small carb — it might have been a 750 CFM [cubic feet per minute], it might have been even smaller — to intentionally limit the horsepower. We knew that particular engine had been on the

The number 71 mule rumbles in the high lane at the Chrysler Proving Grounds five-mile track in Chelsea, Michigan.

dyno so we knew what horsepower it was, and we could very easily calculate the true top speed based on what we got."

Were such precautions really necessary? Chrysler employees who attended the tests recall a number of small airplanes flying over the Chelsea facility — and it's likely that some of the passengers in those planes worked for Ford Motor Company.

Regardless of the slower-than-expected speeds, the tests indicated that this race car was quite different from the Charger 500 and all stock cars that had come before it. John Pointer utilized a unique vantage point to observe the differences.

"They used to have a timing shack right by the guardrail next to the sixth lane," Pointer recalls. "When Glotzbach and Baker went by at 190 MPH three feet from this shack in the Charger 500 they damn near blew it off the foundations! And then the Daytona went by — whoosh!"

Further proof of the Daytona's aerodynamics came during shutdown tests. "There was a little road in the back straightaway, and they told me to shut it off and coast around. Well, I barely made it back to the pit area in the Charger 500," Glotzbach says. "We did the same thing with the winged Dodge, and heck — I had to slide the tires to get it stopped there!"

"Glotzbach turned off the engine coming out of the second turn at Chelsea," remembers Gary Romberg, "and he coasted all the way around the track. That was a five-mile track! And he was still going so fast in front of where our pits were that he spun out in front of the pits just coasting! So we knew that the aerodynamics of the car were good."

The handling of the winged car also received a rave review. "Charlie Glotzbach said you could drive at 180 MPH with one hand," George Wallace says. "He even took Larry Rathgeb for a ride to show him how smooth and easy the thing ran."

Wallace himself ended up being a passenger in the spartan, single-seat interior of the engineering Daytona during tests at Chelsea. As Glotzbach recalls, "They would tie little ribbons or strings all over the hood, tape them on, and George Wallace would ride in the car and watch them with us. He would sit over on the floorboard and wrap his arm around the bar and just sit there. When they put gauges in there he'd sit there and read them, just unconcerned. It didn't bother him any — he was a real cool guy."

Within a week after the first tests at Chelsea, Charlie Glotzbach was orbiting the five-mile track at over 204 MPH in the Charger Daytona.

Buddy Baker says that the car did require some work in those first testing sessions, and that much of it was due to the front spoiler. "We'd never had that much down pressure on the car and the back end was really steering the front," recalls Baker. "We had to tighten the nose down so it didn't push real bad. The car had so much torque — these things

Left: A spoiler mounted 13 inches in back of the leading edge runs the width of the Daytona.

Right: The wing supports are clearly visible as they pass through the trunk area.

had torque like a diesel truck. It was unbelievable! And to develop the front spoiler like we had to have, it took a while to get everything just like you have to have it on the front to balance the car. But once you balanced it, talk about good drivin' — it drove better than your road car to church!"

"They tried different rake angles on the body, or making the rear end higher or lower and changing all that until they came to a happy medium," remembers Glotzbach.

"The angle of the rear spoiler was in the perfect position," says Baker. "It was roof height so you didn't have to run nearly as much spoiler. You'd tilt it like half an inch so you really didn't have any drag whatsoever. It was like having the spoilers of today on the car. Better really, because with the spoiler you don't have anything if you get sideways that straightens the car up. With the wing car it was like an aileron running straight up on both sides, and I mean that thing would steer you right back straight!"

According to Dick Lajoie, however, Baker wasn't always so enamored with the wing. In the early test, when the Daytona wasn't running as fast as expected, "Baker said, 'Take the wing off. That's the problem — the wing is slowing me down!' We said, 'You can't do that. You're going to change the handling!' Well, he got very irate so we took the wing off. And he came around on the warm-up lap and he nailed it when he came by me in the timing shack. And as he went into the first turn you heard a 'burp, burp, burp' and he came coasting back. He was white as a sheet and he said, 'Why don't you put that damn thing back on!'"

The Dodge Charger Daytona was truly a state-of-the-art racing machine. There were no questions about its powerplant — the 426-cubic-inch Hemi was a proven NASCAR winner, developing well over 600 horsepower. It was fed by a single carburetor which gulped air directed by a plenum from slots at the base of the car's windshield. Two small side ducts were added to an enlarged main duct for engine cooling after John Pointer's tests with the mule revealed that the initially proposed use of a 30-square-inch "mail slot" opening on the snout's leading edge was inadequate. The engineers also redesigned front frame members, which were then extended outward to accommodate the radical front end. Each wheel had two shock absorbers, and the suspension featured adjustable torsion bars. Weighing in at 3,900 pounds, the car was 500 pounds heavier than today's Winston Cup stock cars. But naturally, everybody wanted to know about the car's aerodynamics.

Aside from the rounded 18-inch snout, under the nose a five-inch spoiler 51 inches in length was mounted 13 inches in back of the car's leading edge, at a 45-degree angle to the ground. Front end downforce was estimated to be more than 200 pounds. The Daytona also featured streamlined A-pillar wind deflectors. In the back, the wing measured 58 inches across the car and was 7.5 inches wide. It was 23.5 inches above the rear deck, and could be adjusted anywhere from plus two to minus 10 degrees — a total range of 12 degrees. Its surface area was three square feet, and it was shaped like an inverted Clark Y airfoil to create precious downforce — more than 600 pounds worth. Supports for the vertical wing stabilizers ran into the Daytona's trunk area and were connected to the car's chassis on either side of the 22-gallon fuel cell.

Interest in the winged car was so high that Gary Romberg and Bob Marcell delivered a paper entitled "The Aerodynamic Development of the Charger Daytona for Stock Car Competition" to the Society of Automotive Engineers' (SAE) Automotive Engineering Congress in January 1970.

"We had a very unusual SAE conference," recalls Marcell. "Typically when you do an SAE paper you'll get about 25 people in the room to hear about it. Gary and I, when we did ours, we had 500 people in the room! It might have been a record. Of course they had put one of the cars out front, and it had been there for several days and had really built up a

Right: This aerial photograph shows the sprawling five-mile oval at Chelsea, Michigan, where laps above 200 MPH were common in Charger Daytona testing.

Opposite: Dodge's fearless test drivers, Charlie Glotzbach (left) and Buddy Baker (right), perch on the wing of the race engineering Daytona at the Chrysler Proving Grounds.

lot of interest — so we had a roomful. And the first two rows were aerodynamics people from Ford and GM!"

As the tests at Chelsea continued, the speeds climbed and the engineers, designers, and drivers realized just how special the Charger Daytona was. "I knew it right away," says Buddy Baker. "You know quality when you sit down in it, and I've raced in enough good race cars in the past to know that feeling."

The closer the Talladega race grew, the higher the speeds climbed. The fastest of the trap speeds? "I ran 243 MPH at the five-mile oval," says Charlie Glotzbach.

The Dodge Charger Daytona was consistently averaging laps well over 200 MPH — an unheard of speed. But, as Gary Romberg points out, "It was the fastest anybody had been on any track anywhere and we couldn't say anything about it! The racers wanted to set up grandstands at Chelsea and have races there because it was a perfect track for racing!"

Unfortunately, there were no five-mile tracks in the NASCAR Grand National series. But there was a new 2.66-mile track in Alabama. It appeared that when September came, the Dodge teams would strike fear into the hearts of Ford drivers with a new weapon that had been created in record time.

"It's not often that you get to work on something that is that exciting," reflects Bob Marcell. "I mean I work on cars now — I'm general manager of the platform that introduced the Neon and I've got the Viper coupe under me. They're exciting, but hell — even the quick ones take two and a half years before you see things happen. And here we did things in days and weeks and months compared to years. That really is exciting to have that feedback and be able to measure how well you have done and how competitive you are in very short order . . . I still scratch my head and say, 'How in the hell did we do that?'"

The Summer of '69

Buddy Baker was running a Charger Daytona at top speed as he dove off the second turn on the five-mile track at the Chrysler Proving Grounds. So far the Daytona testing had proceeded smoothly, and this run looked to be no exception. Then, roaring into the back straightaway at over 230 MPH, Buddy Baker saw the deer.

"I came off the damn corner over there, running about 235 MPH and I looked up — and here is the biggest deer you've ever seen in your life, right in the back straightaway," Baker recalls. "I made a couple of little turns on the wheel, but whew! When I went by that son of a gun he just kind of bowed up in the back and that's how much I missed him. I was glad to see there was no fur on the windshield. And I don't know who had the biggest eyes when I went by, me or him!"

Despite the occasional wildlife encounter, the Chrysler engineering team and drivers Buddy Baker and Charlie Glotzbach had the Daytona dialed in as a racing vehicle. Runs well over 200 MPH were the norm on the huge Chelsea track, but the mission was far from over. As August 1969 arrived, it was time to take the Daytona south — away from the comfortable five-mile track and onto the NASCAR speedways the car had been designed for.

The first stop was Daytona International Speedway, where Ford's successes had prompted the Charger Daytona's creation. Chrysler Special Vehicles Group's Larry Rathgeb, who was responsible for overseeing the Daytona's development, notes that by the time testing with drivers Glotzbach and Baker began at Daytona, "The car was pretty much all developed as a vehicle. The only thing we had to do was get in hand the way the car handled because of the aerodynamics."

"The car handling at the superspeedways at that speed is 50 percent or more due to the aerodynamics of the car as opposed to the chassis," Rathgeb explains.

"I remember at Daytona when we were first testing the wing car we had enough instrumentation in the car where we knew what Charlie was doing, and he was backing out of it in the first corner," Rathgeb recalls. "He would not go wide open around the track. We said, 'What's the trouble in the first corner, Charlie?'"

"He said, 'Well, when I get in there the bumps are so bad my eyes start to chatter and I've got to close my eyes. I've got to get off the throttle when I close my eyes.' I said, 'Well Charlie, do me a favor — when you go in there and you make your set, even though you close your eyes, don't get off the throttle.' He kind of looked at me and said, 'OK, Larry. I might not bring the car back but that's what I'll do.' And he did. He closed his eyes but he never took his foot off the throttle!"

Opposite: In the wing cars' first appearance in a NASCAR Grand National race, Richard Brickhouse drove a Charger Daytona to victory lane. The win was overshadowed by bizarre circumstances that have become racing legend.

31

Below and opposite: The construction of Bill France's new superspeedway in Talladega, Alabama, was proudly featured in programs from the race track's first season.

The test sessions continued to reveal that all of the homework done at the wind tunnels and the Chelsea track was beginning to pay off. "We would run a lot of short runs but then we'd run a real long one, maybe 100 miles or something like that at Daytona," says test driver Glotzbach. "The car didn't get squirrelly and they pretty much had them figured out when we started racing — there wasn't anything odd about them other than they were fast."

You Bet We're Proud!

We Built It!

Of course, we also build bridges, highways, airports, dams, investigate, design and prepare industrial sites and any general construction project you care to name.

Our pre-stressed and architectural concrete plant is one of the most modern anywhere.

Our engineering staff is the best in the business.

You bet we're proud!

And when you see Alabama International Motor Speedway, you'll see why.

Special Vehicles Group's George Wallace continued to occasionally clamber into one of the wing cars for a 190 MPH observation run with Baker or Glotzbach at the wheel. "It was pretty rough," Wallace recalls of the wild rides. "Before we had much instrumentation that was the only way that you could get any data.

"Typically at Daytona through the turns you'd have about 2 Gs pushing straight down on you and about 1½ Gs pushing horizontally," Wallace says. "If you were trying to write on the clipboard — well, basically it was just impossible. You'd get a bunch of jagged lines.

"We found that if you had your head down as you went into the turn, your neck wasn't strong enough to lift your head and the helmet back up so you could see. If you wanted to watch what was going on, you had to make sure your head was up before you got into the turn. Going at 180-plus MPH with the outside wall six inches away from you is a little interesting."

With the tests at Daytona successfully concluded, the focus of the Chrysler team moved west — to Alabama International Motor Speedway.

NASCAR founder Bill France had already created one palace of speed: Daytona International Speedway. But "Big Bill" wanted to create a second high-banked wonder, one where the racing speeds might even eclipse the hair-raising velocities of Daytona.

France had been considering several potential settings for his dream track, but Anniston, Alabama, resident Bill Ward brought the site that would eventually become the world's fastest superspeedway to France's attention. In 1966 discussions began between France and Talladega, Alabama, city officials to build the track on airport property sold to Talladega by the government after World War II. On May 23, 1968, construction began and the airport and adjacent soybean fields were transformed into Alabama International Motor Speedway.

France's target date was an ambitious one: the 'Bama 400 Grand Touring race was scheduled for

Saturday, September 13, 1969, and the high-profile NASCAR Grand National cars were to race the next day in the first Talladega 500.

It was at this new and untested superspeedway that the Dodge Charger Daytona would make its competition debut. France's target date was shared by the Chrysler engineers who wanted to have the new car fully deployed among the Chrysler teams in Grand National racing when Talladega opened. With the inaugural 500-mile race just over a month away, a sense of urgency surrounded the Chrysler team. Testing began in August 1969 in the heart and heat of Alabama. Earlier in the year a Ford Torino Talladega had won the 1969 Daytona 500. Chrysler was determined that a Daytona would win at Talladega.

Rumors that all was not well at Talladega had spread through the Grand National world. In May 1969 paving had yet to begin at the 2.66-mile track, cutting into the amount of time that the tire companies would have to prepare suitable race tires for the anticipated record speeds. In late July NASCAR Grand National racer Bobby Allison had cruised the track in a passenger car, reporting that the now-paved surface of the huge track was "extremely rough."

The first week of August, Ford drivers LeeRoy Yarbrough and Donnie Allison whipped Torino Talladega stock cars around the track. The sleek Fords that raced in 1969 were Torinos with a special aerodynamic snout, and though the Talladega model had been named after the just-completed superspeedway, the track showed the cars little gratitude. After racing through the 33 degrees of banking at over 190 MPH, both Ford drivers reported it to be a bone-jarring experience.

The Dodge Charger Daytona first rolled onto the surface of Alabama International Motor Speedway in the third week of August. "There were screw-ups when they built the speedway," says Chrysler test engineer John Pointer. "It was intended to be a clone of Daytona — a little bit bigger and a little bit steeper bank. But Alabama down there is clay. It's not sand and I guess they had some torrential rains just before they paved the track and right in the middle of the turns there were some waves.

"At 200 MPH lap speeds at Talladega in the corners the driver is experiencing almost 2.5 Gs down in his seat and very close to 2 Gs lateral — this isn't chopped liver to experience," Pointer continues. With the high speeds and the uneven track surface, "The car was being bounced up and down about three inches at about 40 times a second."

Here are some construction scenes from the new Alabama International Motor Speedway, showing the heavy equipment of the Moss-Thornton Construction Company used for the building of the $2.66 mile super speedway.

(1) During the early days of the construction project, these huge road builders and earth moving machines were used for construction of the 6,000 foot jet airport strip and then the speedway itself.
(2) Placing the asphalt surface over the packed ground and gravel base.
(3) The International Speedway Corporation's planning committee for the Talladega project. Standing left to right are Leo Spanuello of Union/Pure Oil Company; Aubrey Vincent, counsel for the speedway organization, and William C. France, Vice President and General Manager of International Speedway Corporation. Seated left to right are William H. G. (Bill) France, president of the organization; and Claude Brineger, President of the Union 76 Division of Union Oil Company of California.
(4) Construction stages at the speedway, showing huge crane with cable attached to machine for leveling surface on high 33-degree banks.

The regular Chrysler test drivers — Buddy Baker and Charlie Glotzbach — were augmented at Talladega by Bobby Isaac, a Dodge driver who had three Grand National victories and 27 top-five finishes in 1968. All three drivers emerged from their Charger Daytona runs with complaints about blurred vision and physical discomfort. Larry Rathgeb, who oversaw the test runs, investigated what was going on inside the cars.

"Our instrumentation people were out of Huntsville," remembers Special Vehicles Group coordinator George Wallace. "This was where we ran our race garage at the time because Huntsville is a whole lot closer to the NASCAR tracks than Detroit is, and it kept the race cars out of the mainstream of Detroit operations.

"The instrumentation people had all worked in the Chrysler/NASA projects. The NASA stuff was cutting back at the time, so we had a lot of good people available."

Among those called in to help was Bill Wright, an aerospace engineer with the Chrysler Corporation's Defense and Space Division. At the Talladega test, Wright remembered that during the research leading up to the Apollo space program, astronauts in test vehicles had complained of temporary loss of sight, nausea, chest pains, and other ailments. The problem was eventually attributed to vibrations from the rocket engines.

"We had accelerometers in the cars," Wallace explains, "and they started measuring the frequency of the vertical data. They found that it was very similar to the vibration period that had been discovered in some of the early Apollo tests. They found that whatever the frequency was it had a very definite psychological effect on the astronauts — just a general feeling that things weren't right. And the supposition was that this was what was happening to the race drivers."

This condition became known as the Pogo Effect. But the effect of the track on the drivers wasn't the only problem the Chrysler team discovered in tests at Talladega. "They'd done a trick paving job intended to abrade as the cars raced so [the track] didn't build up rubber," John Pointer notes. "That just chewed the life out of the tires. They could get about five laps at speed out of the Goodyears."

But no matter how strange the events leading up to the debut race week at Alabama International Motor Speedway were, they paled in comparison to what happened once the race teams arrived in Alabama for the running of the Talladega 500.

On August 14, a group of 11 Grand National drivers had met to form the Professional Drivers Association (PDA) and began to actively recruit members. By the time the Talladega race came around, it appeared that Alabama International Motor Speedway could very well be the site of a showdown between the drivers in the PDA and Bill France. The PDA was demanding certain benefits and changes in the structure of Grand National racing, but Bill France did not like being told what to do.

The tension between the PDA and France led to concerns that the race itself might become the target of a walkout. Those in the know wondered whether top drivers like Richard Petty, PDA president, would even take to the track for qualifying. Still, PDA or no PDA, the Chrysler team was committed to getting their wing cars in a position to win the Talladega 500 — if it was held.

Although Larry Rathgeb headed up all of the engineering work and testing on the Charger Daytona, the racing program was administered by Product Planning under manager Ronnie Householder. Rathgeb was concerned that the new Daytona parts and aerodynamic characteristics might confuse some of the Dodge teams who had hurriedly converted their Charger 500s into wing cars. It was quite conceivable that this unfamiliarity could lead to a public relations disaster — a car other than the Charger Daytona might be the fastest qualifier of the Talladega 500.

"Owners hadn't seen it, and very few of the drivers had ever driven it," Rathgeb remembers. His solution was to enter a Chrysler-owned Daytona — the actual engineering test car.

"When it came time for the race I talked to Ronnie Householder," Rathgeb continues. "I asked him if it would be all right if I entered the car in order to have it at the track at the time of the race, so that some of the owners and drivers could look at the car and get a better idea as to how to set their cars up. He said, 'Sure, see Ray Nichels and he'll give you a number and set it all up and go ahead and enter the car.' So I did. I entered the car."

The engineering car was considered a Ray Nichels Engineering entry for the Talladega 500 — car number 88. Nichels's Indiana operation was used as the central clearing house for parts for the entire Chrysler racing program, so the deal was simple to arrange. But Rathgeb wasn't satisfied with just having the car at Talladega.

"The car was there at Talladega and I was having some problems with Ronnie trying to get the car run," Rathgeb says. "He did not want the car run on the racetrack. I said, 'Well, Ronnie, nobody is running their car very fast. The Ford cars are faster than we are, and I know this car is capable of a lot better speeds than the other cars and I've got to get it on the racetrack.'

"They had a regulation there that if the car had not run at least two laps in practice it could not be qualified and could not run in the race. So he said, 'All right, put Charlie in it and send it out for a couple of laps — but make sure that he doesn't go any faster than 185 MPH!'"

The results of that Tuesday September 9, 1969, practice run were predictable.

"I put Charlie in it," Rathgeb remembers. "I said, 'Now look, Charlie, Ronnie says this car is not to go any faster than 185 MPH.' And he looked up and smiled, and he said, 'Sure, Larry.' And he went out and he did three laps. He went out on a warm-up lap, and then he went fast, and the next lap was 199 MPH. It was unbelievable! The whole place came apart, just came apart. When he came in, I waved

The race engineering Daytona was Larry Rathgeb's pride and joy. It got him in plenty of hot water during qualifying for the race at Talladega.

The Dodge Sheriff was a popular character in Chrysler advertising. Here he discusses strategy with (from left) Bobby Allison, Bobby Isaac, Charlie Glotzbach, and Buddy Baker, but not even the law could mend the rift between the Professional Drivers Association and Bill France.

him into the spot and we covered the car and just left it and just sat there and worried. Then Ronnie called me. He called me everything! He told me he'd have my job, I'd be in the street — oh God, it was terrible!"

Rathgeb's next plan for the blue number 88 car was to put it on the pole. "We had to qualify that car in order to get the pole," Rathgeb believes. But to do that, especially with an angry Ronnie Householder already fuming about Glotzbach's hot lap speeds that had included one circuit at 199.987 MPH, Rathgeb had to come up with a plan.

"We had that car and we had the practice laps," Rathgeb explains. "Time was getting shorter and people were going to go out and qualify and I needed to qualify that car. I didn't want to talk to Householder, so I called my own boss. That was Paul Bruns, who commanded the engineering group at Chrysler. I told him what the story was, what had happened at the practice sessions, and that the car was now prepared and could qualify and we needed to put somebody in it to qualify."

With the blessing of Bruns, Rathgeb got Charlie Glotzbach to qualify the 88 Daytona — no small feat considering Glotzbach was under the control of Householder and was scheduled to drive Ray Nichels's other entry. But Glotzbach had just begun to drive for Nichels Engineering; his first race for Ray Nichels's team had been only weeks earlier at the Yankee 600 in Michigan on August 17, 1969. Glotzbach had been testing for Rathgeb for much longer.

"I went to talk to Charlie," Rathgeb recounts. "I said, 'Charlie, I understand what your situation is but when it comes to qualifying, would you qualify this car?' And he said, 'Of course I will.' I said, 'Well, what's Nichels going to do?' He said, 'I don't know what Nichels is going to do, I don't know what Ronnie's going to do, I don't care about my job — I just want to qualify that car.' And that's what he did."

Charlie Glotzbach braved the harsh surface of the track for first-round qualifying on September 10 and guided the engineering Charger Daytona to the pole position of the Talladega 500 with yet another fearsome lap speed — 199.466 MPH.

"The problem was that our stable of race car people didn't have all of the latest sheet metal and all of the tricks that we had been developing both on the track and in the wind tunnel," recalls Chrysler aerodynamicist Gary Romberg. "We were trying to get all of that technology into all of the race cars, but it wasn't there yet. So Rathgeb made the decision to take the car that was developed on the racetrack — the engineering car — and enter it in the race.

"Glotzbach jumps in the car and goes 199 MPH — he's the fastest qualifier. Well, that makes all of us technical people very happy because we knew the car was fast. But the politics explode both inside and outside of the company. Our whole policy with the racers was that you give them the technology, but Chrysler was not supposed to enter a car. Good God! Nobody did that! There was a big witch hunt, and they didn't have to hunt very far, because Rathgeb was right in the frying pan."

Larry Rathgeb's plan to get Glotzbach in his engineering car and on the pole had worked. But there had been an earlier plot that was much different and even stranger, one with elements that might be more at home in a James Bond novel rather than a stock car tale.

Late one night in the week before the Talladega 500, a street version of the Charger Daytona cruised beneath the black Alabama skies. Inside were Larry Rathgeb and Special Vehicle Group's George Wallace. This clandestine journey had been set in motion by Rathgeb's call to his boss at Chrysler, Paul Bruns.

Rathgeb says that his call to Bruns took a surprising turn. "I figured that I really had to qualify that car. I told him that Charlie Glotzbach was working for Nichels and he was under Nichels's command and, therefore, Householder's command. And he said, 'Well, what do you want to do?' I said, 'I'd like to put Richard Petty in the car.'"

Petty had traditionally been one of the Chrysler Corporation's most ardent supporters and most successful Grand National drivers. His blue number 43 Plymouth stock cars consistently ended up in victory lanes across the country. But in late 1968 Petty had become upset when he learned that no revisions were planned to the Plymouth cars, when the Dodge drivers would be racing in heavily redesigned Chargers.

Rather than drive what he considered to be a non-competitive car, Richard and his crew chief Dale Inman went to Ford for the 1969 season, signing a one-year deal to compete in a sleek Torino Talladega. The idea that Larry Rathgeb would make such an overture to a driver signed to another factory was surprising enough; that Paul Bruns would give him the official okay was even more astonishing.

"He said, 'Go ahead and do it,'" Rathgeb recalls of Bruns's reaction. "And I said, 'Well, he's a Ford racer . . .' 'I know what he is! If you think that's what you'd like to do then go ahead and do it.' I said, 'Okay, thank you. Will you stand behind me?' And he said, 'Sure, I'll stand behind you.' I called

Richard from the motel and I said, 'Richard, I really have to talk to you personally.' And he said, 'Okay, come on over.'"

"Whatever town we were staying in, Richard was staying somewhere else," George Wallace remembers. "We had a road-going Daytona . . . and we drove over to where Richard was staying and we talked to Dale Inman first. Richard had already gone to bed, so we woke him up. We had our engineering test car that Ray Nichels was going to have somebody drive, and we had happened to throw in the back a set of the number 43 decals."

With tensions mounting daily between the PDA and Bill France, it was beginning to appear likely that Richard would never take to the track in his Torino. The tension between the PDA and France was the focus of Rathgeb's sales pitch.

"I said, 'Richard, I know you're going to back out of this thing with the Ford. That's obviously going to happen. But everybody's going to go do this qualifying. Since you're not going to qualify the Ford, would you qualify this Dodge for me?' He said, 'Oh, I would really love to do that, but I have an obligation to Ford. Whether I run or not, I really can't run your car.'"

"He was quite diplomatic," George Wallace recalls of Petty, "but he said he was contractually obligated to drive a Ford. But we did make the offer and, though we would have been greatly surprised if he had accepted it . . . we would certainly have been prepared to follow through on it."

Wallace dismisses the fact that having Petty qualify the Charger Daytona in its debut would have made for great publicity. "We were more interested in seeing what he could do with the car," Wallace maintains. "Our interests were more that he was the best driver around and we'd like to have him in our car."

While all of these sub-plots were being played out, there was one constant — Alabama International Motor Speedway was displaying a ravenous appetite for tires. The blinding speeds were being set in short burst runs, but even in sessions of fewer than five laps the race tires were showing signs of stress and failure. Both Firestone and Goodyear were scrambling to solve the problem, but the difficulties appeared insurmountable.

As if the tire problem wasn't trouble enough, drivers were climbing from their cars disoriented and winded by the vibrations of the Pogo Effect. Some admitted they were scared by the potential for disaster.

On Friday, September 12, 1969, Charlie Glotzbach — using Goodyear tires — and Donnie Allison — on Firestones — each ran a brief tire test to check out the companies' latest attempts at a compound designed to tame the rough track. Donnie blistered one set of Firestones after a mere four laps, changed tires and had the same thing happen after four more laps. Firestone withdrew its tires from competition. Glotzbach's Goodyear test was cut short before tire wear could be accurately determined when his Daytona broke its right front A-frame, but it's safe to say Goodyear was as worried as Firestone.

Chrysler's Gary Romberg and Dick Lajoie had driven from Michigan down to Talladega and, upon their arrival at Alabama International Motor Speedway, were greeted with one of the lighter moments of what had been a very anxious week. "Gary and I drove into the racetrack," remembers Lajoie, "and here's a sheriff's car. He'd bought a Daytona and he'd put the gumball light up on the wing!"

On Saturday, September 13, 1969, there were no light moments as tensions mounted in the garage area.

With the walkout looming, Dodge public relations director Frank Wylie crossed paths with Richard Petty. "I exchanged some words with Richard Petty on the subject," Wylie says. "I said, 'You can walk away from this — if you don't run the race it doesn't make any difference. But there are a hell of a lot of guys out in the back end of the field, and it makes a lot of difference to them. Wendell Scott, and people like that. The big money guys are scaring the little money guys out. You make your point if you don't run, but those guys could run and they could make some money.' But he was adamant about it." Despite the fact that less well-financed racers like Scott needed the prize money offered in the Talladega 500, they trusted Petty to act in their best interest as the PDA's leader.

Bill France was watching the first Grand National race at his dream track turn into a nightmare, and he blamed Richard Petty and the PDA for most of his problems. France had gone so far as to threaten to enter a car in the Talladega 500 to gain entry to the PDA, and did take a few laps in a stock car to "prove" the track was safe. France drove a Holman-Moody Ford numbered 53, which had been used to compete in United States Auto Club (USAC) events and was not NASCAR Grand National-legal. Unimpressed, Petty and the PDA drivers were unyielding in their opinion that the racetrack was unsafe. The fact that the 400-mile 'Bama 400 Grand Touring support race was run that afternoon without incident proved nothing, as the Grand Touring cars ran at slower speeds than the Grand National competitors.

The final showdown at Talladega. Richard Petty (wearing sunglasses in center of group) puts his foot down for the PDA while Bill France (wearing a dark cap with white stripe) insists that there will be a race on Sunday.

Charlie Glotzbach was all smiles when he saw his new Charger Daytona, but he didn't get to drive the car in its first NASCAR race.

Finally, late that afternoon, the voice of Bill France himself came over the garage public address system: "All those who are not going to race, leave the garage area so those who are going to race can work on their cars." The first truck engine that started was that of the Petty Enterprises team, hauling its Torino Talladega away from the track. Petty's was the first of thirty Grand National teams to head for home.

Drivers and car owners had met with manufacturer representatives before making the decision to leave. Were the Chrysler officials understanding when Charlie Glotzbach informed them he would not be driving his new Daytona on Sunday?

"I think that they were because they knew what was going on," Glotzbach now says. "We ran Firestone tires and we ran Goodyear tires and you couldn't run at the speeds we would run. You couldn't run five laps without tearing the tires up. I believe that they understood perfectly what was going on and they didn't demand that we go out there and race.

"I sat on the pole, but the circumstances with the track conditions and everything just took something away," Glotzbach continues. "The track just wasn't safe. Somebody would have gotten hurt if we'd been out there racing like we should have been racing. They brought in a bunch of cars from everywhere to run because a lot of us took our cars and went home."

Bill France was forced to pad the field of the Talladega 500 with more than twenty drivers from the Grand Touring division which had raced Saturday afternoon in cars like Camaros and Mustangs. Of the Dodge factory teams, only two cars remained: Nord Krauskopf's orange number 71 K&K Insurance Daytona driven by Bobby Isaac, and the Nichels Engineering purple number 99 Daytona, which Glotzbach had originally been scheduled to drive.

Bobby Isaac had kept away from the PDA, feeling insulted that he wasn't trusted enough to be invited to the very first PDA meeting on August 14, 1969. For that reason, the fact that Isaac stayed to compete in the Talladega 500 was never held against him.

And why had Ray Nichels decided to let his car compete? "I think Nichels Engineering felt obligated to do it," reflects Charlie Glotzbach. "At that time they were building all of the Chrysler cars for all of the teams."

The driver's seat of the Nichels Daytona was filled by young Richard Brickhouse, who spent the days before the race agonizing over the decision of whether or not to race. Having joined the PDA just before the Talladega race, he risked the wrath of the drivers who walked out if he decided to stay and race. But 29-year-old Brickhouse faced other pressures — pressures directed at him by Chrysler racing boss Ronnie Householder.

"They called me at home and said, 'You're going to drive a factory car when you get here,'" Brickhouse recalls. Up until then Brickhouse had been struggling to break into the Grand National ranks, and this looked to be his big opportunity. Then he got to Talladega.

"I arrived in the midst of all this hoopla," Brickhouse says. "It was unbelievable. There were so many forces at work, there was so much controversy. The Chrysler people, the Ford people, the Firestone people, the Goodyear people, Bill France — everything was up in the air and there was nothing but confusion."

When the PDA walked out on France, "that just added fuel to the flame. Everybody was jockeying around wondering what was the best thing to do. Everybody was trying to cover their own you-know-what. We were trying to do the same thing."

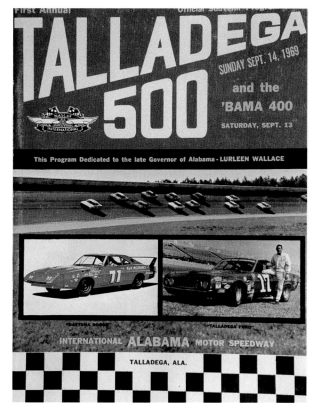

Right: The Daytona was too new for a photograph of it to appear on the cover of the race program.

Left: The only two wing cars to take the green flag at Talladega were the Daytonas of Bobby Isaac (number 71) and Richard Brickhouse (number 99).

The public story has always been that Brickhouse made the decision to race of his own free will, despite having joined the PDA the week before. But Brickhouse, struggling to become a factory driver for Chrysler, had more than free will helping him make his decision to race. Years later, Brickhouse still hesitates when speaking of the pressures of that chaotic weekend.

"If you want to be frank about it, I remember Ronnie Householder walking up to me in the garage area that morning," Brickhouse says. "He was the head of Chrysler racing, and he pointed at the car and he said, 'Either you run this car or somebody else will.' And those were his very words. That was coming from my boss man.

"I was committed to Chrysler. I had worked too hard, we were running out of money, and I knew if I didn't make it that was going to be it. It was the fellows on the fringes, like me, who stood to lose or gain. I was under Chrysler's thumb, and whatever they said to do was what I was going to do. And they made it clear that I was going to drive that car."

As the director of Dodge public relations, Frank Wylie decided who would drive for which Dodge teams. The PDA's strike had thrown Dodge's plans into chaos, but Wylie had a job to do — he had to make sure the Daytona won its first time out. To do that, he had to help Householder convince Richard Brickhouse to drive.

"Brickhouse wasn't sure that he wanted to get into it," Wylie says. "It took a little talking. I just thought it was important to win that race. I was not in favor of the strike."

Brickhouse, who had tried to personally inform Richard Petty of his resignation from the PDA, was sought out by Wylie, who wanted Brickhouse to formally announce his resignation so that he could drive the Daytona. Brickhouse's resignation was read over the track's public address system.

The stage was finally set for the Talladega 500. There would be two winged Dodge Charger Daytonas in a ragged thirty-six-car field of Mustangs, Camaros, and independent Grand National cars. Bobby Isaac would start from the pole, even though Charlie Glotzbach, LeeRoy Yarbrough, Cale Yarborough, Buddy Baker, and Richard Petty had all recorded quicker laps during first round qualifying on September 10. Everyone who had qualified faster than Isaac's Daytona had left with the PDA.

On Sunday, September 14, 1969, the green flag waved and the Talladega 500 was finally underway. After months of exhausting work, Larry Rathgeb

Richard Brickhouse was relieved to reach victory lane, but it was a hollow victory for Chrysler's Special Vehicles Group. And, on top of it all, Jim Vandiver claimed he had actually won the race.

watched his treasured Dodge Charger Daytonas thunder through a field of under-powered competition. Rathgeb had dreamed of his Daytonas battling side-by-side with the best of the Ford teams. What he saw was something else entirely.

"I had tears in my eyes for the first lap," Rathgeb says quietly. "Those two cars out in front and nothing but a bunch of scrap behind them. It was a heartbreaker, to work that long and that hard on that kind of a program — and then have it just fall apart when it should be in its glory . . . It was a terrible heartbreaker."

Bobby Isaac led the first lap before experiencing problems with the number 71 Daytona, and eventually finished fourth. That left it up to Richard Brickhouse to salvage what little glory was left at Alabama International Motor Speedway.

The first thing Brickhouse had to do was learn what driving a Daytona was like. "I didn't get any practice in that car," Brickhouse points out. "But I had plenty of time — I had 500 miles!" Fortunately Brickhouse was a fast learner.

"I paced myself — you've got to save your car to win," Brickhouse says of that day. "I never turned the car loose until the end of the race. They had the scoreboard mixed up and I thought I was a lap down, but I wasn't. I ran it wide open then."

Brickhouse began turning laps so fast that a crewman was dispatched with a pit board to try to get Brickhouse to ease off the pace before a tire failure wrecked the car. Brickhouse remembers the crewman ran all the way to the edge of the racetrack. "He came clean across pit road out onto the grass and he had 'SLOW DOWN!' written [on the pit board]. They had clocked me at 197 MPH. I don't remember how many laps I ran like that, but I thought I was behind."

Richard Brickhouse's purple number 99 Daytona won the Talladega 500 by seven seconds over Jim Vandiver in Ray Fox's Charger 500 at an average speed of 153.779 MPH. Sixty-four thousand spectators had witnessed one of the strangest NASCAR races ever held. Tire wear was excessive although there were no crashes, and race speeds varied between 170 and 185 MPH under green flag conditions as the drivers tried to find a speed safe for competition. Brickhouse had discovered that the higher the lane he ran in, the less tire wear he had, so he planted his Daytona up close to the wall on the 33-degree banking — a strategy that paid off with a win for the wing car in its first outing.

Unbelievably, though, controversy was not finished with Talladega. Jim Vandiver was convinced he had won the race. The record books show he did lead the most laps in his Charger 500, 102 of the 188 that made up the 500-mile race. Vandiver maintained that Brickhouse's seven-second margin of victory was actually how close Vandiver was to lapping Brickhouse, and that the scoring mix-up Brickhouse referred to was not a mix-up at all.

The program cover for Darlington's 1969 Southern 500 depicted an exciting battle between David Pearson's Ford and Bobby Isaac's Daytona even though no wing cars raced in the event that year.

Crew chief Harry Hyde (left) and driver Bobby Isaac discuss wing angles in the duo's quest to be the first Chrysler team to beat the Fords in head-to-head competition.

"Oh, I won the race — they hoodoo'ed me out of it," Vandiver still maintains. "I was driving for Ray Fox, and after the race he argued with them for two hours on the scoring. They could never disprove what we were saying. But you know, you can't do anything with NASCAR when it comes to something like that. Every time we pitted, we pitted on a caution flag. Brickhouse pitted twice on a green flag and we lapped him. There wasn't any doubt about it. In my mind there is no doubt we won the race.

"It's just one of those things — it was a deal where it was the first time they had run that car and they wanted that car to win," Vandiver says of the Daytona. Ironically, Vandiver later went on to race successfully in his own winged Dodge.

Not surprisingly, Richard Brickhouse disagrees with Vandiver's claim. "One scoreboard was reading one thing and one was reading another," Brickhouse explains. "They had a man up there working the thing manually, standing up there changing the numbers, and one read one thing and one had the other. There was some confusion, but there's no way he could have gotten a lap on me — I was 9, 10 MPH faster than he was."

Richard Brickhouse's victory was proclaimed official, and the Talladega 500 finally made its troubled way into the record books.

Fortunately for all concerned, the remainder of the 1969 season was less controversial. The PDA drivers raced at the next event on the Grand National schedule in Columbia, South Carolina, although the Charger Daytonas — designed to compete only on the largest NASCAR tracks — would not be used again until the Grand National circuit raced in Charlotte, North Carolina.

Meanwhile, testing continued and Chrysler's team was faced with the decision of when to use the exotic Daytona. "We were going to run at Rockingham, North Carolina," remembers Larry Rathgeb. There was a race at the 1.017-mile track toward the end of the season. "Householder said, 'We're going to run the standard cars at Rockingham — don't use the wing cars. They're expensive and we don't want to hurt them.'

"I said, 'No, Ronnie, we should use the wing cars.' And he said, 'Damn it, the aerodynamics aren't going to help you at Rockingham.' And I said, 'Ronnie, it is going to help. Let us go down there and test. We're going to test there anyway. Let us test there with the wing car, the engineering car, and bring in one of the other guys with their standard car.'

"And so he had Mario Rossi come down," Rathgeb continues. Rossi was the owner of the Dodges driven by Bobby Allison. "Allison was driving and they could not go as fast as the wing cars. The same driver in the same car, and we even switched engines in the cars. There was no way the standard car could run as fast as the wing car, even on a one-mile track. You had this improved downforce on the car, and the handling was so much better in the corners." Rathgeb had proven the wing cars had benefits even on the one-mile tracks.

Soon the Charger Daytonas, Torino Talladegas, and Cyclone Spoiler IIs were pulling into Charlotte

Motor Speedway for the National 500 on October 12, 1969 — the second race for the winged Dodges.

"It scares me just to look at it," noted Ford star David Pearson of the ominous wing car, but on that Sunday a Ford conquered the Daytonas at the 1.5-mile speedway. Donnie Allison drove his Torino Talladega to victory lane, with the Daytonas of Bobby Allison, Buddy Baker, and Charlie Glotzbach taking second, third, and fourth. This performance was not what the Chrysler Special Vehicles Group had in mind. Again, tires were the culprit.

Fords had qualified in the first four positions for the race. The fastest qualifier, Cale Yarborough, said of the Daytonas, "We had looked at those cars and wondered quite a bit about how we would stack up. It's a great feeling to win the pole under these circumstances."

Yarborough now says of the Daytonas, "That was a drastic change in car design when they came out with the winged cars, but it sure proved to be

Racing tire technology could not match the sophistication of the wing cars, and tire failures kept the Daytonas from showing their full potential. Here, Buddy Baker has just received a fresh set of tires for Cotton Owens's Dodge.

successful. They were like airplanes, and they were certainly built to race. They were good race cars."

When the Charlotte race began, the Daytonas dominated until the left rear tires began to break up in chunks. When the tires were fresh the Daytonas went anywhere their drivers pointed them, but after 12 to 15 laps the cars became tremendously loose — almost undrivable. It was all the Dodge men could do to hang on to their top-five finishing positions.

"I didn't realize that the aerodynamic design of this race car was that far ahead of our current tire designs," Ronnie Householder fretted. "Apparently this is the case."

Larry Rathgeb and Charlie Glotzbach were immediately dispatched once again to North Carolina to test the Daytona prior to the October 26, 1969, 500-mile race at Rockingham, but the additional testing laps did not help.

Although Charlie Glotzbach put his wing car on the pole of the American 500, LeeRoy Yarbrough and David Pearson put their Fords in first and second to win the race. Buddy Baker and Dave Marcis — both in Daytonas — managed third and fourth. Bobby Allison had his Daytona in second place when he crashed on the 20th lap. Richard Brickhouse's Daytona wound up 31st with suspension problems.

The last superspeedway race of the Grand National season would be December 7, 1969, at Texas International Speedway, a brand new two-mile racetrack with turns banked at 22 degrees. A strong showing by the Daytonas was absolutely essential to save face, even though the problems encountered had been almost entirely tire-related. But the public didn't care about tire problems — all they cared about was who won the race.

Dodge was encouraged by the new speedway's smooth track surface, universally hailed by nearly all of the Grand National drivers as being "like velvet." Things began to look even better when Buddy Baker put his dark orange Daytona on the pole with a qualifying lap of 176.284 MPH.

The Dodge gang was all smiles as the Texas 500 began and Baker set sail. He led for 149 laps and was a lap ahead of the nearest Ford competitors. Then the unthinkable happened. With just over 20 laps remaining and the field circling the track under caution, Baker looked to his pit slot instead of where he was going and rammed right into the Daytona of James Hylton. Baker fell from the race, his car's pointed snout now crushed inward. The Dodge people were in shock, Baker's team owner Cotton Owens was furious, and a Ford was in first place.

Donnie Allison's Talladega led the way when the green flag came out, but Bobby Isaac's Daytona soon pulled into the lead. The Ford then got back by the wing car, with Isaac running close behind as the laps dwindled. Suddenly Allison's right front tire failed, and Isaac tenaciously hung on to the

Under this Daytona's hood was the command Dodge hoped everyone would follow.

lead until the checkered flag waved. At last, the Daytonas had won in head-to-head competition with the Talladegas.

The turbulent 1969 season had drawn to a close. The 1970 NASCAR Grand National season would clearly be a pivotal one. There was every reason to believe that the tire companies would have enough time to prepare tires capable of handling the much greater downforce of the wing cars.

Ford was expected to have a revised body in 1970, and many of their drivers were already worried that it would not be as aerodynamic as the proven Torino Talladega configuration. The Dodge Charger Daytona drivers anxiously awaited the 1970 Daytona 500.

But there was one unknown variable in the Grand National equation. Plymouth had decided to build a wing car of its own.

At the end of the 1969 season, this was the Dodge stock car racing factory team.

Building a Better Bird

"They said, 'Will you come back with Chrysler if we build you a wing car?' And I said, 'Let me see what you've got.'"

That's how Richard Petty remembers the events that led to the creation of the second Chrysler wing car, the Plymouth SuperBird. Petty had left Plymouth for Ford for the 1969 Grand National season, and Plymouth wanted him back. If they had to build him his very own Plymouth wing car, then that's what they would do — Petty was that valuable to the racing program.

The words Petty and Plymouth had become synonymous during a spellbinding 1967 season, when Petty drove his Plymouth to an astonishing 27 wins in 48 starts. He finished in the top five 38 times, and won three times as much money as his closest competition. Long tracks, short tracks, paved tracks, dirt tracks — it made no difference. "King Richard" won at every track where the Grand National cars raced.

But in 1968, the win total dropped to 16 — still extremely respectable, yet Petty's performance over the 49-race season found him behind David Pearson's Ford and Bobby Isaac's Dodge in the season-long points tally.

When Petty got wind of the Dodge division's plans to create the Charger 500 for their racers to drive in 1969, Petty grew anxious, wondering what he could expect for his Plymouth team. Perhaps a new, more aerodynamic version of the Plymouth could help him regain Grand National domination.

"Along about the middle of 1968 we found out they were going to have a Dodge Charger 500," Petty remembers. "We saw the pictures and we said, 'Man! Look at that — that is great!' And I said, 'Well, what's the Plymouth going to look like?' And they said, 'The Plymouth's going to look like it usually does.'"

One can imagine Petty's disappointment. To stand still in the face of Grand National racing evolution is to lose ground — fast. The thought of racing against Charger 500s and Torino Talladegas in 1969 while driving the same Plymouth he was running in 1968 had The King squirming on his throne. Since he technically drove for Chrysler Corporation, Petty thought it made perfect sense for Chrysler to reassign him to compete in a Dodge Charger 500.

"I said, 'Okay. Are you going to give me a Dodge to drive?'" Petty recalls. "And they said, 'No, you're still going to drive the Plymouth.' And I said, 'Ain't no way I can drive a stock Plymouth against that thing.' And they said, 'We're not set up where we can get a new Plymouth right now.' So I said, 'Just go ahead and give me a Dodge.' And they said, 'No, we can't do that.' I said, 'I'm not going to do it. I'm not going to run against that deal there.'

"I said, 'I'll go over and talk to the Ford people,' because they had been romancing me pretty heavily

Opposite: A highly stylized artist's conception of the Plymouth SuperBird.

since we blew them all away in 1967," Petty continues. "Plymouth said, 'All right, go on.' So that's what we did, basically. We just walked across the street and talked to Ford and they said, 'Sure. No sweat.' They had come up with the new Ford Talladega, the sweptback job. I didn't even go back to Plymouth. They had already told me they were not going to operate with what I thought I could survive with, so we signed up with Ford for a year."

Chrysler Special Vehicles Group's George Wallace explains. "Richard's main objection was to Ronnie Householder," says Wallace. "He did not like the way Householder was running the effort."

Any questions about how successful Petty might be in a Ford were answered when he went to Riverside, California, and won the first Grand National race of 1969. He won nine more times that season, but more importantly, he damaged Chrysler's image by defecting to Ford. The Plymouth people paid Richard a visit at mid-season.

"In June, they came along and wanted to know what my contract with Ford was," Petty says. "I said that it was just a year contract. They offered me a winged Plymouth. I imagine they had already done some work on the thing because they already had the experience from the Dodge. It was a big job and an expensive job but they decided that's what they would do and we decided to go along with it."

With a commitment from Petty to return for the 1970 season, work began at full speed to get the SuperBird built. Some at Plymouth thought it was simply a matter of having Creative Industries — which had built the street Daytonas — tack the sloped nose and the tall wing onto a bunch of Plymouth Road Runners. Unfortunately, creating the second winged car wasn't so simple.

"In June of 1969 Plymouth committed to do a race car," aerodynamicist Gary Romberg recalls. "And what they wanted to hear from us was, 'Oh, it's already developed, all you have to do is get in line at Creative Industries and you can do SuperBirds.' And that wasn't true. They said, 'When we're done we'll have a car that's just as good as the Daytona.' And we told them time and time again, 'That isn't the case. You won't be as good.'

"There was a lot more track work to bring in the SuperBird than there was in the Daytona. We had to flog the SuperBird; it was flogged both inside by John Pointer and the rest of us, and it was flogged outside by Petty."

Dick Lajoie, another Chrysler aerodynamicist, says, "One of the problems with the SuperBird was that when styling got involved we were at the wind tunnel in Wichita. They did not like the leading edge of the nose as low as the Daytona. So they asked us to raise it about an inch and a half, two inches, and that increased the drag. Then there was a problem with the backlight. We couldn't bring the backlight out as far. They just wanted to put a plug and kind of bubble the backlight. That increased the lift and drag."

These problems with the nose and rear window area underscore a fundamental difference between the Daytona program and the SuperBird program. With the Charger Daytona, the aerodynamics and engineering people were shielded from all outside influences. With the SuperBird, the styling department at Plymouth would get to have their say, and after what they saw in the Daytona, they had a lot to say.

"I was horrified by it when I first saw it," says John Herlitz of the Daytona. Herlitz was manager of the Plymouth intermediate car studio. He had worked on the Satellite and Road Runner, and the SuperBird would be in his domain. Herlitz was no fan of the Daytona's unusual appearance. "The extreme length of the nose was such an exaggeration," Herlitz critiques. "To me, it was a ridiculous-looking car, particularly with the tail fins up and canted out to the sides."

"We knew we were on the right track if styling thought it was world-class ugly," says aerodynamicist Romberg mischievously.

But when it came time to begin designing the SuperBird, styling was involved from the beginning. "The racing guys came to me and they asked if I could

Look closely in the back seat area of the engineering SuperBird undergoing high speed testing and you'll see George Wallace bravely clinging to the roll cage.

go out to Creative Industries and take a look at a proposal that they had put together," recalls Herlitz. "We spent a whole day out there deciding what could and couldn't be done, trying to give it at least a little commercial appeal."

Commercial appeal was even more important with the SuperBird than the Daytona. While Dodge had been required to make 500 street Daytonas to go racing, NASCAR had since changed the rules. The new car count required was either 1,000 cars or a number equal to one-half of the total number of a brand's dealers — whichever was higher. Plymouth would have to build four times as many SuperBirds as Dodge had Daytonas, and these highly modified Road Runners would have to be built before January 1, 1970.

"The way the racing engineers wanted to do it at first was to use the Plymouth body up to the front fenders," continues Herlitz. "Then they were going to put the Charger front end on it. I said, 'Guys, you can't do that,' because all of the lines from the Dodge fender are going to be running into a Plymouth door that has no lines in it. So at that point they grudgingly agreed to a unique nose cone for the Plymouth.

"Once we got over that hurdle, then we began to work cooperatively to get a solution that was serviceable for the performance ends and commercial requirements."

SuperBird construction presented unique challenges. The fenders of the Road Runner were not suitable for having a nose cone attached. Instead, an awkward transplant of Dodge Coronet fenders was necessary, along with a special plug to allow for a smooth hood-to-nose cone transition. The hood itself was a re-worked Coronet piece. The rear backlight would also have special glass and a plug, but, unlike the similarly altered Charger 500 and Daytona, the SuperBird would have a vinyl roof hiding the surgery — a concession to cost in light of the number of street SuperBirds that had to be made.

Was Herlitz happy with the way the car eventually appeared? "Well, I was as happy as I could be. It was a functional solution and there were things

One method of tracking air flow around the SuperBird was to study the movement of yarn attached to the car's body during testing.

that we tried to do, such as using the Fury park and turn lamps in the lower grille area. Also the large black graphics on the upper part of the nose cone surrounding the pop-up headlamps — that gave some interest and tended to shorten the apparent length of the front overhang. So those were a few of the little tools that we used to try and make it more commercial."

To make the car faster, extensive wind tunnel research was conducted. Dick Lajoie remembers "camping out" at Wichita State University while studying the ⅜-scale models. The aerodynamicists were searching for a way to reduce drag much as they had with the Daytona.

"We went through the same thing but it was a tougher job," says Bob Marcell, who was heavily involved in the SuperBird effort. "When you start changing the backlight area you can only go so far, or it gets extremely expensive. If you start changing the trunk and changing the quarter panels and the roof, by the time you get done you've almost redone the car. We could change the base of the backlight, but because we couldn't change the touch-down points on the side of the car and the roof, we could never get the SuperBird to be quite as good as the Daytona was."

"They wanted it to be cheaper so they didn't want us to go into the 'Dutchman' — which is a piece at the very base of the backlight," Gary Romberg explains. "So we couldn't pull the backlight out and make it into a fastback — which was worth about two or three or four percent in drag. But they said, 'No. That costs too much.' John Herlitz asked, 'Can we at least make those tail fins a little bit bigger?' We ran some numbers in aerodynamics, John Pointer did some testing at the track, and it looked like it was going to increase the stability just a tad. So we said yes."

George Wallace reveals that originally the engineering department was worried that the large vertical fins on the Daytona might stabilize the car too much and make it hard to turn. "But after seeing how well the Charger Daytona worked, for the SuperBird we made them bigger and went to about 100 percent of the possible aerodynamic stability. The SuperBird was a little more stable, but it did have more drag."

In the end, the stabilizers were approximately 40 percent larger than those on the Daytona, and they were tilted inward more. They also had a more dramatic sweep in profile. The SuperBird was 221 inches long — nearly 18.5 feet. As on the Daytona, the A-pillars were streamlined. Up front, the air inlet had been redesigned to prevent overheating, and the SuperBird's beak cut into the wind at a slightly higher angle than the Daytona's did.

Although June 1969 is regarded as the beginning of the SuperBird program — or the "Belvedere Daytona program" as it was referred to in some 1969 Chrysler memos — the work got off to a fitful start. In an April 1970 memo detailing the SuperBird's development history to engineering's H. Paul Bruns, the Aerodynamics Group complained that in August 1969, Plymouth suddenly decided to drop the whole wing car program. Plymouth refused to accept the changes to the car demanded by the group's aerodynamic evaluation due to costs and the time required to make the revisions.

Two weeks later, though, Plymouth changed its mind. The program was on again, with the backlight compromises and a less costly nose design that fell just short of the Aerodynamics Group's ideal recommendation. The big problem was that two precious weeks of development time had been lost.

Working under a strict deadline, the SuperBird team forged ahead. "You kind of got instant feedback," Bob Marcell notes. "When we finally decided what shape the car wanted to be, we went to Creative Industries. We went over there with templates we had made in ⅜-scale, we blew them up to full size, and in a couple of days we had clayed the wing and the front end and the backlight onto the car.

"Within a couple of weeks we were in a position to actually have parts off soft tools. We could build actual cars and have them running at Chelsea. A lot of that stuff happened very quickly."

The spartan interior of the SuperBird Grand National race car.

With the specifications to start production now in place, Larry Rathgeb began to test the SuperBird race cars. "The problem we had with that car was getting it turned," Richard Petty recalls. "Most cars up to that time, you could turn the car but then the car would get loose and feel so free in the back that maybe you'd have to let off. The problem we had with the SuperBird was that when you'd get to the end of the straightaway you'd be running so fast, you'd turn it, and you had so much overhang in the front that the car wouldn't turn down into the corner. We had the big spoilers on the front, and we used to do trick stuff with the spoilers and everything on the front of the car to get the car to turn.

"The rear, you never had any trouble," Petty continues. "You had that wing and you'd just tilt that thing any way you wanted to and nail that thing to the ground. It was really an easy car to drive — you could really make it comfortable."

Petty credits the Chrysler engineers with helping Petty Enterprises make the transition to the SuperBirds in preparation for the 1970 season. "They were a lot of help," Petty says. "They'd run the things in the wind tunnels and they could tell us things to do. A lot of the stuff they would tell you to do you had to go out and try, because they couldn't say, 'This will fix it.' It was still trial and error. But they were a good crowd of people to work with. Once we got into the fold, you had one set of engineers and they engineered the Dodges and the Plymouths — they were Chrysler people."

The Chrysler team knew that Petty had a different approach than most of the drivers they had been working with. "Most of the race car people were the old test pilot types," says Gary Romberg. "They'd just go out and run really fast and turn left and win. That was the old approach. But Petty would say, 'Let's find out what makes things happen and how they go fast.' Those were the best years, the relationships with Plymouth. When you talk to him he'll say it was like family, and it was."

Petty was more than pleased with the work of Chrysler engineers. "They went through every combination that was possible, whether it was different springs, different shocks, different sway bars, different weights — and they put it in a book," Petty says. "It must have been about a 2' x 3' book, and it had every combination that you could put under that car. It would tell you what the percentages were and what the car was supposed to do. Now this was done before they had computers to do it — somebody had to figure this out by hand. So it was a heck of an engineering force Chrysler had. They really did a super job."

Of course, as pleased as Petty was with the new SuperBird he would be driving and the support he received from Chrysler's engineers, he was equally happy with another aspect of his new deal with Plymouth — Petty Enterprises replaced Nichels Engineering as the clearinghouse for all of Chrysler's rac-

ing parts, cars, and equipment. Chrysler's racing headquarters moved from Highland, Indiana, to Level Cross, North Carolina.

"In order to get Petty Enterprises back, the wing car was just part of it," Petty acknowledges. "We wound up being the distributor of everything Chrysler did. We had every piece that it took to build a race car. If we didn't have it we would build it. So you want a race car? We'd go in there and wipe the surface plate off and start bringing in the metals and the jigs. We had the jigs to mount the doors, to mount the rollbars, to mount the A-frames, to mount the shocks. All of that stuff was done like an assembly line deal. We had one manufacturing building where they built the cars, and we had our race shop just backed up to it. Anything Chrysler sold in the racing line, if it was sold to a sports car team in Peru, South America, it came through Petty Enterprises."

Clearly Plymouth and Chrysler had made a tremendous investment in their relationship with Richard Petty. Still, the public's focus was glued to Petty's performance on the racetrack. After all of the wind tunnel work, after all of the engineering work, after extensive testing at the Chrysler Proving Grounds in Chelsea, after more testing at Daytona International Speedway, Plymouth's faith in Petty Enterprises and the new SuperBird would be rewarded. Surprisingly though, it was not Richard Petty who delivered NASCAR Grand National racing's biggest prize to the Chrysler camp.

A finished work of racing art. Richard Petty's Plymouth SuperBird as it appears today in the Richard Petty Museum in Level Cross, North Carolina.

Aero Wars

The new decade began with two aerodynamic wing cars poised for combat in the NASCAR Grand National arena — the proven Dodge Charger Daytona, winner of the final 1969 superspeedway race, and the brand new Plymouth SuperBird.

The wing cars were generally not used at smaller racetracks. These exotic secret weapons were designed to be raced when the stakes were highest — on the towering banks of the very fastest superspeedways. But there was one racetrack with elements of both short tracks and superspeedways — the nine-turn, 2.62-mile road course at Riverside International Raceway in Riverside, California. It was here, on January 18, that the first battle of the 1970 aero wars would be fought.

Richard Petty was back with Plymouth, and Petty Enterprises was fielding two new SuperBirds at Riverside. Of course, Petty had driven the SuperBird in testing sessions in the weeks before the team headed west to California for the Motor Trend 500. But cruising around a superspeedway all by yourself was one thing — racing in close quarters with 41 other Grand National cars was something else entirely. During practice sessions, Petty realized he had a problem with his 18.5-foot wing car.

"It was the first time we ran the car. We got there, and the thing was so long with that nose that you couldn't see — it just kept on going!" Petty laughs. "So I sent the boys down to a local auto parts place and they bought us a radio antenna. We put it on the front spoiler and we ran it up where I could see it sticking out there so I could tell how far that thing was out there! Once I drove it in one race, I had some depth perception then.

"But to begin with, I didn't know how close I could get behind anybody or not. But I had that little radio aerial out there and I learned that it was out there four or five feet. Then we took it off and didn't run with it anywhere else, but that was how I learned how far that nose was out there."

At Riverside, the other Petty Enterprises SuperBird was to be driven by the popular California driver Dan Gurney, who had won this event five times previously. Although officially listed as a Petty entry, the Gurney car was really the Plymouth engineering SuperBird making a competition appearance as the engineering Charger Daytona had in the week before the first Talladega race. And, like the engineering Daytona at Talladega, the dark blue number 42 engineering SuperBird became the fastest qualifier with a lap of 112.060 MPH. Parnelli Jones's quicker time in the Wood Brothers' Mercury was disallowed by NASCAR because the car used an ineligible type of Firestone tire.

Opposite: Garry Hill's painting of the 1970 Daytona 500 shows the wing cars of Pete Hamilton (40), Buddy Baker (6), Bobby Allison (22), Charlie Glotzbach (99), Bobby Isaac (71), and Richard Petty (43).

Bobby Allison at the wheel of Mario Rossi's Dodge Charger Daytona.

The SuperBirds showed promise during the Motor Trend 500, but wily A.J. Foyt used his skill, experience, and a Ford to win the race. USAC competitor Roger McCluskey, who drove one of the new SuperBirds, finished three seconds behind Foyt to take second. Petty and Gurney took the checkered flag in fifth and sixth.

All in all the wing car performances at Riverside were respectable, but the cars weren't really designed to turn left and right, since road course events were a rarity on the Grand National circuit. The most important test of the wing cars was just weeks away — the 1970 Daytona 500.

Gurney's presence as a Petty Enterprises driver had been a one-race deal, and at Daytona the second Petty-blue SuperBird would be driven by Massachusetts driver Pete Hamilton. "They wanted two front-line Plymouths," recalls Petty, "So they came to us and asked, 'Will you take on another car?' I said, 'Okay, we'll do that. Now, you're wanting us to do this, you're paying the bills and doing the whole deal, who do you want as a driver?'

"They gave us a list of four or five drivers," Petty continues, "and as far as we were concerned Pete Hamilton was head and shoulders above anybody else they had on there. Even though he'd never won a Grand National race, he'd won Rookie of the Year and he'd run good from time to time. He'd won other championships and we thought he was a capable driver."

"I was ready to compete as a Grand National driver," says Hamilton. "I had gotten to know the engineering people at Chrysler, and I got to know the tire engineering people at Goodyear and Firestone, and I think that helped me. I attended the University of Maine for a year in engineering, and I was intensely interested in the mechanical part of the car. Richard knew that I had done my fair share of winning on the short tracks and he saw that I worked on the car, which is what he did."

Hamilton's first runs in a Grand National car were in a Charger 500 prepared by Jim Ruggles. "I had run about 163 MPH in a Sportsman car at Daytona," Hamilton says. "I got in that Charger and the first lap I ran was 187 MPH. So that was 24 MPH faster than I had ever been — on the first lap! So that was a big change. I just knew the first time I made that lap I was just committing hari-kari. I knew I

Left: Petty Enterprises' arsenal heads into combat.

Right: Young Pete Hamilton — king of the superspeedways in 1970 — in a Plymouth public relations photo.

was either going to make that lap or die. That was the atmosphere." Clearly Hamilton had the will to win by any means necessary.

"I think that Richard saw some abilities, saw some desire, a fire to win, and he knew that I was capable of working on the car and knowledgeable about the car," Hamilton says. "I think that type of driver is important to him. So I got a call from Richard Petty to come and drive his second car. It wouldn't have made any difference if it wasn't a SuperBird — it would have made no difference if it was a Jeep Cherokee."

Although Hamilton had been at Riverside with the Petty team, his duties were limited to helping prepare the Gurney car. "I worked on the car while Gurney drove, because I knew nothing about a road course." Hamilton explains. "Then we went on to Daytona."

When the Petty Enterprises team pulled into Daytona International Speedway in February 1970, Pete Hamilton had yet to drive the SuperBird he would be racing in.

"The first time I ever drove a wing car was at Daytona," Hamilton admits. "I never tested in it. We went to Daytona, unloaded the beast, turned it on, and put it to the floor in high gear and saw what was going to happen. That was it — no testing."

Pete Hamilton placed sixth in the first 125-mile qualifying race, one position better than Richard Petty. Under the complicated Daytona qualifying procedures, that meant Hamilton would start ninth in the 500. Evidence of the expected Ford-Chrysler showdown could be seen in the front row of the starting grid — Cale Yarborough's Wood Brothers Mercury was on the pole with a run of 194.015 MPH, but right next to him was the second-fastest qualifier, Buddy Baker, in his Cotton Owens Charger Daytona.

When the drivers in the 1970 Daytona 500 took the green flag at the 2.5-mile superspeedway to begin the race, Pete Hamilton drove the only way he knew how — flat-out.

"I was a lot younger and crazier than Richard, and we always ran at the big places a lot faster," Hamilton explains. "I don't mean that in an insulting way . . . Richard had five more years of experience and knew that he could get hurt on a superspeedway."

"Pete Hamilton had no idea that he could get hurt — he was young and stupid," Hamilton says of himself. " 'There is no way I could get hurt here' — I was absolutely fearless. That is a big advantage to have in a driver at a big joint. Not only did I not have any fear of getting hurt in the car but I had no fear of hanging my ass out and sticking the car in the fence,

because I just did what I thought I was supposed to do. That makes an absolutely treacherous race car driver — if he finishes."

The experienced Grand National drivers weren't at all familiar with the aggressive Hamilton, and Hamilton used that fact to his advantage in the Daytona 500. "They were clueless, and if you throw your car sideways in front of an old-timer it would scare him to death," Hamilton says. "When you have a book on a guy, you trust him and want to run close to him. When you don't have a book on a guy — whew! You don't want to get near this wild man! That's a big advantage, when the other drivers don't have a book on you. I had that card, and I played that card."

Just seven laps into the biggest NASCAR race of the year, Richard Petty's engine blew in the number 43 SuperBird. Most observers expected to see Hamilton's blue number 40 wing car pull down pit road so Petty could take over the driver seat of the other Petty Enterprises entry. Petty, however, had other ideas.

"I didn't run but a few laps before I blew a motor," Petty recalls. "So that eliminated us, and then when I came into the pits everybody really anticipated me getting in Pete's car. I said, 'Hey — this is not a second-line car, this is a first-line car. This is Pete Hamilton's car. He's going to do the best that he can with it.'"

Hamilton was indeed doing his best, and by lap 63 his SuperBird was in first place until Bobby Isaac stole the position in his winged Daytona. Hamilton, despite charging hard, drove a smart race and was in a good position as the laps dwindled. On lap 187, Dick Brooks's SuperBird suffered an engine

Above: The Petty Enterprises crew jumps towards Richard Petty's SuperBird as Charlie Glotzbach guides the Nichels Engineering Daytona down pit road.

Opposite: The Daytonas and SuperBirds have David Pearson's lone Torino Talladega surrounded.

Following pages: Buddy Baker (6) and Charlie Glotzbach (99), two of the swiftest members of the Dodge Scat Pack.

failure and the final caution flag of the day came out. David Pearson, whose Holman-Moody Ford had been extremely tough all day, was the leader and pitted quickly for two tires. Hamilton guided his SuperBird down pit road and the Petty Enterprises crew provided him with four fresh tires. The green flag came out with nine laps remaining — just 22.5 more miles to race around the banking to win stock car racing's biggest prize. But first Hamilton had to get by crafty Pearson, who was known as "The Silver Fox."

"When they dropped that green flag, I never let off," Hamilton remembers. "Earlier in the race I had been letting off, but now I never let off the gas and the car somehow made it. Pearson was first, I was second, and Bobby Allison was ahead of me and I knew he was a lap down. I remember pulling up beside him going through turns one and two — he was in the second lane and I remember being right up against the wall. He kind of sensed that and he let me by because he knew he was a lap down."

Now Pete Hamilton had nothing in his sights but the rear of Pearson's blue and gold Torino Talladega. "That put me close to David, close enough to him," Hamilton says. "I got up there and I passed him, passed him on the bottom."

That put Hamilton in the lead, but the best place to be at the end of a superspeedway race is often second — where a slingshot pass can be set up on the final lap. "I was so knowledge-less about the draft," Hamilton admits. "As soon as I passed him he dropped in behind me. The realization came to me that he was where he needed to be to win the race, in second, not in first. So off we went."

It had all come down to this. Ford versus Chrysler. The famous Holman-Moody Ford team against the legendary Petty Enterprises. The Torino Talladega against the SuperBird. The cagey veteran

Images from the glory days of Chrysler's NASCAR program. Bobby Allison leads Pete Hamilton and Buddy Baker through a turn (opposite), while Baker and Richard Petty battle side-by-side (left).

against the fearless young driver on 31 degrees of treacherous banking. It was everything the greatest stock car race of them all — the Daytona 500 — should be.

"When he was in the draft he could get down the straightaway better than us but he wasn't that good in the corner," Hamilton remembers of Pearson's late-race charges. "He took a couple of stabs at me. He made one good shot at me and came up underneath me and got beside me.

"I remember holding him tight, and I remember coming off the fourth turn and seeing him kind of dropping back — and he never got another chance. The race was over. I never let off for a lap after the checkered flag. I ran that car just as hard as it would go for another lap — because I wasn't taking any chances!"

Pete Hamilton won the 1970 Daytona 500, and the SuperBird lived up to its name. For the first time since 1966, the most important NASCAR race, and the invaluable bragging rights that came along with it, belonged to Chrysler.

But just because they had won at Daytona didn't mean that the work of Chrysler's Special Vehicles Group was over. The Chrysler team wanted to make sure they knew why they had won at Daytona.

"They took the spoiler that I had on that Plymouth, the front spoiler, and they took the rear wing and put them in the wind tunnel, because for some reason that car ran especially well," says Pete Hamilton.

Hamilton occasionally joined the dynamic duo of testing, Charlie Glotzbach and Buddy Baker, in on-track sessions. That also meant getting used to the wing car veterans' high jinks. "There was Baker and Glotzbach and Larry Rathgeb and Gary Romberg and a whole slew of us," Hamilton remembers. "It became a contest between Baker and Glotzbach.

"They'd come in and they'd say 'Take out three rounds of bite,' which makes the car looser. And then Glotzbach would drive the car and he'd say loosen it up a little. Well, they got that thing so loose nobody could drive it. They could not run it at any

Action along pit road in 1970. Bobby Allison's Daytona (opposite) and Pete Hamilton's SuperBird (left) get fuel and tires. New tire designs held up much better than the tires used in 1969, allowing the wing cars to fully demonstrate their superiority.

speed at all — I mean I got in the car, had them tighten it up six rounds, and wound up running faster than the both of them. They did it because it was fun running the car sideways through the corner."

Special Vehicles Group's George Wallace, who used to ride in the engineering Daytona with Glotzbach or Baker to observe various aspects of the car's performance, astounded Hamilton when he did the same thing during SuperBird tests.

"Some of these goofy bastards would get in the car at Talladega with us and run 190 MPH around there," Hamilton marvels. "George would take a handkerchief out and hold it inside the car to see what the wind was doing! This guy trusted me enough where he's wrapped up back where the back seat would be with his legs and arms and ass hung around the roll cage to see what the hell the air was doing inside the car. I knew that those guys were all a lot sicker than us. Holy cow! It's unbelievable. It was a wild time, and I feel very fortunate to have been involved with that."

The next race in which the wing cars competed was the Carolina 500 in Rockingham, North Carolina, on March 8. The 1.017-mile North Carolina Motor Speedway came as a pleasant surprise to the

There were no speed limits on pit road in 1970, and the wing cars would soar in and roar out in a cloud of tire smoke.

Chrysler wing car teams. As George Wallace explains, "At Daytona and Talladega the wing cars were fast like they should be, but they really worked on the medium speed tracks because of the tail fins. At places like Rockingham the extra stability of the tail fin really helped."

The other half of the Petty Enterprises team was victorious at Rockingham, when Richard Petty — despite spinning twice — beat Cale Yarborough's Mercury and Dick Brooks's SuperBird. Bobby Allison's Daytona, which had qualified on the pole at 139.048 MPH, was fourth, followed by Pete Hamilton in the other Petty car.

"I worked as hard on Richard's cars as I did on mine, and he on mine as he did on his," Hamilton explains. "There were 18 or so of us at Petty's and we all worked really hard on the cars to make them last and to make them finish. It was done very much as a team. Even though at the big joints I was winning and running faster, Richard seemed to handle that well — and he'd beat me at places like Rockingham and Atlanta that required a little more finesse and more intelligent driving than what I possessed at that moment."

While much attention was focused on the SuperBirds' victories at Daytona and Rockingham, Bobby Allison struck a blow for the Daytonas the next time the wings raced when he came from a lap down in his red and gold car to win at the 1.522-mile Atlanta International Raceway on March 29. Again, Cale Yarborough had to settle for second and Pete Hamilton claimed third. Charlie Glotzbach was the fastest qualifier with a lap of 159.929 MPH.

With such strong competition between the two manufacturers, it was inevitable that both Ford and Chrysler might search for an extra edge — even if that edge wasn't quite legal under NASCAR's rules.

Chrysler's Bob Marcell used his experience as an amateur photographer to study the Ford competition. "We knew if you could lower the car, that would make a tremendous difference," Marcell says. "I did a lot of 35 mm photographs which we blew up after the race and looked at the cars. It was very obvious that Ford was able to get their cars lowered during the race. That was pretty significant – you could pick up more than a few miles per hour just by lowering the car. Ford was finding a way through their suspension system to raise and lower the car so that they could still pass inspection after the race."

Not that the Ford teams were the only ones. According to highly placed sources at Chrysler, one of the tricks used by Chrysler teams involved placing a plastic cup over the adjusting bolt of the front torsion bar. Under the forces of the car at speed, the plastic cup would disintegrate. Then the adjusting bolt would move up about 3/16 of an inch, and the front end of the car would lower up to a full inch. Once that trick was discovered by NASCAR's inspectors, the clever rule-benders moved to the rear of the car where they rigged up a method using slotted plates which would shift the car downward at speed.

When NASCAR began measuring ride height of the stock cars both before and after races, the Chrysler racers had to be more ingenious. Using a complex system of oil and compressed air in a special hydraulic device that was concealed in the depths of the car, some drivers were able to lower the wing cars to race and then raise them back up for inspection.

Such innovation is often referred to as "a creative interpretation of the rules" in the racing world. But no matter what it's called, both sides enthusiastically indulged in these practices with varying degrees of success.

The SuperBirds made their first appearance, along with the veteran Daytonas, at the repaved 2.66-mile Alabama International Motor Speedway in Talladega on April 12, 1970, and again Pete Hamilton was victorious in a superspeedway race. Bobby Isaac,

"Buddy Baker — America's No. 1 Charger" reads the bumper sticker attached to the crushed rear end of the Cotton Owens Daytona. On this day Baker charged a little too hard.

who had started from the pole after qualifying at 199.658 MPH, followed in the orange K&K Insurance Daytona.

"You have to run a little bit sideways anyway," Hamilton says of superspeedway racing in a SuperBird. "You'd have to get into a slip angle kind of thing in the corner to really keep the RPM up, but they would hold that.

"The SuperBirds seemed to be a little bit more stable because of the vertical stabilizer. It's laying against the air. When you pull out next to somebody there's not that much air to lay against. When a car would get to a certain degree of yaw, by moving that aerodynamic center of pressure to the rear, you could almost lay the car against the wind bank. The more yaw you got into, the more wind bank you had which

Pete Hamilton prepares to win again in superspeedway competition.

Darlington Raceway in Darlington, South Carolina, hosted the aero wars on May 9, 1970, and this time a Ford beat out the winged competition as David Pearson drove the Holman-Moody Ford to an easy victory. Richard Petty, who had wrecked his SuperBird in practice, suffered one of the most violent crashes in stock car history in his back-up Road Runner. Amazingly, Petty was not seriously hurt. Dick Brooks, 1969 Rookie of the Year, raced his white SuperBird to the runner-up position, and Bobby Isaac's Daytona placed third. Glotzbach had again been fastest qualifier, punching out a 153.822 MPH lap around the 1.366-mile track.

would straighten the car out. We tested at Talladega and the engineers would tell us you couldn't spin the car out — that wasn't true. But you could certainly run a looser car, which was speed. You could throw the car in the corner more."

The Talladega race also revealed one of the more dangerous aspects of the wing cars. When Buddy Baker's Daytona experienced a left front tire failure, pieces of the ruined tire battered the oil cooler mounted inside the car's fender. Twenty-four quarts of hot oil began to spill out of the dry sump oil system. When the oil ignited, Baker was in an inferno.

"It sprayed hot oil on the exhaust, and I don't have to tell you what happened then," Baker says. "When I got it stopped the inside of the car was on fire. Somebody had miscued and put enamel paint on the inside of the car and that made it almost like a fuel. I realized I had to calm down to get my helmet off to get out because the strap around my neck had tied up into the shoulder harness.

"As soon as I realized I was going to burn up if I didn't calm down I just sat down there in the fire, took my helmet off and stepped out of the window. But up until that time I was just trying to get any kind of fresh air, and I'd jump almost to the fresh air and the shoulder harness would jerk me back inside. I think it was probably the scaredest I've ever been."

Dick Brooks is one of the few Grand National drivers who raced both in Daytonas and SuperBirds. According to Brooks, the Dodge was a little slicker. "It had less rear-quarter panel. As far as driving, I don't think there was that much difference. The wing car didn't buffet around at all. It stayed so straight. When you turned that corner and pointed that car where you wanted to go, that's where it went. It was just an awesome car."

Brooks soon learned some of the aerodynamic tricks of his SuperBird. "If a guy in front of you were to get a little out of shape or something, you could kind of ease down underneath it a little bit and kind of tuck the rear back under them," Dick says. "There'd be four or five of us hooked together nose to tail and you'd stick that nose up underneath the other guy's bumper and if he started slipping around a little you could straighten him out. You could also take the car right out from under him.

"I know one time we were practicing at Talladega and there were five or six of us running," Brooks continues. "I lost it going into the third turn — it just got away from me and I'm literally dirt-tracking across this turn, absolutely sideways. And Bobby Allison was the next car behind me and he just kept going down and going down — he pulled me on in. We finished off that lap and came back in. I got out of the car and Bobby came walking over and he said, 'Well, you owe me one now!' And he was right."

At Charlotte Motor Speedway on May 24, Bobby's brother Donnie drove his reddish-orange number 27 Banjo Matthews Ford to victory in the World 600. Cale Yarborough beat Benny Parsons's Ford for second place, with Tiny Lund's Daytona the only Chrysler in the top five. Fred Lorenzen, one of NASCAR's greatest drivers, had been retired for three years but chose this race to make his comeback — in a Charger Daytona. Engine problems left Lorenzen with a 24th place finish. The pole position had gone to Bobby Isaac, who blasted off a lap of 159.227 MPH at the 1.5-mile speedway.

On June 7 the wings rode into Michigan International Speedway for the Motor State 400. Cale Yarborough rode out with the win in his white and red Wood Brothers Mercury, barely beating Pete Hamilton to the line. Several teams — both Ford and Chrysler — claimed that Yarborough should not have been scored on the lead lap, but the victory stood. Hamilton was fastest qualifier, touring the two-mile track at 162.737 MPH.

Jim Vandiver, who had been involved in a scoring controversy of his own at the debut race at Talladega, drove his blue and yellow Dodge Charger Daytona to sixth place, one of several strong runs Vandiver had enjoyed since switching to the wing cars.

"That was the best driving race car I've ever driven," Vandiver says enthusiastically. "Let me tell you what happened at Talladega. Me and Cale Yarborough hooked up and I was drafting him and we passed about ten cars in the backstretch. We got to the third turn and the car got loose — it got all

Charlie Glotzbach in victory lane, the home of many wing car drivers throughout the 1970 Grand National season.

Donnie Allison's Ford leads the cars in the outer lane while Joe Frasson's Daytona is first in the lower groove.

the way crossed up. I'll never forget it because I was so far sideways I was looking straight down at the grass in the doggone third turn and I said, 'I know I'm going to spin out.' And the thing got all the way sideways, and came back! I just kept fighting and fooling with it and it came all the way back. And if it had been any other car it would have spun out. I said, 'Man! I can't believe this!' It is by far the best-driving race car I ever drove."

Chrysler's George Wallace believes that for some drivers, the wing cars were a little too good.

"One of the reasons Petty was never that happy was because the car was too good," Wallace says. "It made other drivers too good. The car would help them catch it if the tail end started to get away from them. Richard was good enough that he could let the tail end hang out and hold it there — he was probably the best driver in the business. The other people, if that car started to get away, they'd be into the wall. These cars made other people as good as Richard and understandably that would not make him happy."

But Petty was happy on June 14, when the Grand National cars returned to Riverside, California, and Petty put his SuperBird into victory lane at the Falstaff 400. Pole-sitter Bobby Allison piloted Mario Rossi's Daytona to second, with James Hylton's Ford a distant third.

Petty agrees with Wallace's assessment of the wing car stability. "It was all but impossible to spin one of them out at speed," Petty says. "You'd see these cats when they'd blow motors or have trouble when they'd blow a tire, you'd see these cars get completely sideways. But before long the front would come back. When they'd get around there that wind would just catch that rudder on the side and stabilize that thing on top. I don't know that I've ever seen one completely spin out."

One Grand National driver returned to the garage in an agitated state, proclaiming that the Hand of God had reached down and straightened out his Daytona when he had cut a tire. Possibly, but as Richard Brickhouse notes, "The faster they'd run the more stable they got."

Of course, Charlie Glotzbach had long since been sold on the Daytona's stability. "You could turn loose of the steering wheel and it would straighten out by itself," Glotzbach says. "That's what Larry Rathgeb said it would do and it did. It's amazing. Hard to believe, really."

The Daytonas and SuperBirds returned to one of the superspeedways which they were designed to run on for the Firecracker 400 at Daytona International Speedway on the Fourth of July. This time though, the fireworks went to Ford when Cale Yarborough captured the pole with a lap of 191.640 MPH and Donnie Allison claimed another victory in the Banjo Matthews Ford. The rest of the top five consisted of wing cars: Buddy Baker, Bobby Allison, and Charlie Glotzbach in their Daytonas, and Dick Brooks at the wheel of his SuperBird.

The wings then flew north — to the 1.5-mile Trenton Speedway in New Jersey for the Schaefer 300. This race track was dominated by the wing cars during the July 12 race, with Richard Petty winning again, despite crashing his SuperBird in practice. Bobby Allison and Charlie Glotzbach finished ahead of Dick Brooks. James Hylton's fifth-place finish was the best Ford could muster. Bobby Isaac started yet another race on the pole after he clicked off a 131.749 MPH lap.

Dick Bown raced the only wing car entered in the next three races, risking the relatively fragile nose of Mike Ober's SuperBird in Tennessee short track events in Bristol, Maryville, and Nashville. His best finish was sixth place in the East Tennessee 200 at Smoky Mountain Raceway.

The all-out aero conflict returned for real in the Dixie 500 on August 2 at Atlanta International Raceway. The weekend began with Fred Lorenzen's Daytona grabbing the pole after a lap of 157.625 MPH. When the race ended, Richard Petty was ahead of the Ford factory cars. This time Cale Yarborough and LeeRoy Yarbrough followed the blue number 43 SuperBird to the checkered flag.

One of the strengths of Richard Petty and the entire Chrysler effort was their "cookbook," a collection of data gathered by the Special Vehicles Group's Larry Rathgeb that gave teams a base line setup at any track they visited.

"What happened was when you got ready to go to Charlotte, the people at Chrysler would say, 'Okay, this is the recommended setup,'" Petty says. "Nobody probably ever ran that exact setup, but that was a safe setup and you could get around the racetrack. Then you would make minute changes because every driver was a little different. But this was a good, safe setup that they could give you up front."

One key chapter of the cookbook concerned the recipe for setting the angle of the wing on the Charger Daytonas and SuperBirds. Larry Rathgeb figured out a clever way to calculate the proper angle to achieve the most advantageous downforce.

"Through our experiments, mostly at Daytona Beach where you can run wide open, we found exactly what angle the airfoil wanted to give the best

Opposite: Harry Hyde, the innovative and clever crew chief behind Bobby Isaac's charge to the 1970 Grand National championship.

performance," says Rathgeb. "And then we found out in the wind tunnel what that downforce was. So we went to each of the racetracks, and for each of the tracks we took the qualifying speeds and then found the wing angle that it would take to get the same downforce that we had at Daytona Beach.

"And we had a curve made, and we'd go to the racetracks with a little inclinometer that I'd bought at Sears. We bought a whole bunch of them and distributed them. You'd set it on top of the wing, and then you'd move the wing up until you achieved this angle. You needed to know what the attitude of the wing was to the car as opposed to the world. So you just took this thing and put it on the sill, read the angle, then put it on the top, subtracted the two and you came up with the proper angle for the wing. It didn't matter whether it was flat where the car was — uphill, downhill, it made no difference."

Those who tried to improve the cookbook's recipes sometimes wound up with indigestion. "Buddy Baker was driving for Cotton Owens but had wrecked the car, and the next race was at Texas International Speedway," recalls John Pointer. "By the time they got done straightening out the sheet metal they said, 'How are we going to set up the car?' Well, they said, 'We'll start with the cookbook.' They just threw the parts in there, raced out to Texas, backed the car off the trailer, shazam! They're well above the lap record, faster than they've ever gone.

"Cotton said, 'Boy, there's a lot more where that came from!' and he started tweaking the car — he's going to make it faster. And how much did they slow it down? Eight miles per hour! Finally, they're getting desperate — it's getting time to qualify. They'd changed so many things they didn't really know what was causing the problem so they said, 'The heck with it, we'll just go back to the cookbook.' Bang! They were right back where they had been."

Only one wing car showed up at the .437-mile track at Ona, West Virginia, on August 11. Buddy Baker was behind the wheel of Neil Castles's Dodge Daytona when the West Virginia 300 got underway at International Raceway Park. Why would Baker run a car designed for massive superspeedways on such a tiny track?

"Let me tell you, if that car had had a bumper on the front it would have raced anywhere," Baker says. "There was that aluminum nose on it and the next thing to hit was the radiator, so that was the only reason we didn't race them everywhere. They absolutely had the body style to race on a short track or anywhere else. If you could have just put a nice metal bumper all the way around the front, that would have been the car to race everywhere."

Baker's run at Ona only lasted seven laps before brake problems sent the Daytona into the garage.

On a return to Michigan International Speedway for the Yankee 400 on August 16, the wing brigade flew to the top six positions. Charlie Glotzbach led the way in his purple number 99 Daytona, followed by Bobby Allison, Dick Brooks, Bobby Isaac, Pete Hamilton, and Buddy Baker. The race at the two-mile track was especially sweet for Glotzbach since he was also fastest qualifier at 157.363 MPH.

Isaac's fourth place finish, coupled with his consistently good performances throughout the season, put him in the position to make a run at the Grand National championship. The mastermind behind Isaac's K&K Insurance Daytona was Harry Hyde, who had become one of the most respected and innovative crew chiefs in Grand National racing. Hyde relied heavily on the partnership between his race team and the Special Vehicles Group engineers.

"We'd go out to Lockheed and we'd have Harry out there and Petty's crew," aerodynamicist Dick Lajoie recalls. "Old Harry would say, 'Now you guys are giving me all of the information you're giving to Richard, aren't you?'"

"The Chrysler engineers were very, very smart," Hyde says. "I don't think I've ever been around a group of people any finer than those engineers at Chrysler. They worked with us. They were really in-

Four wings demonstrate precision flying.

terested in what we thought and what we thought would help. When they found out something we wanted, they'd come back and fix it for us or come back and test it for us, tell us if it was right or wrong."

Harry Hyde has seen many NASCAR stock cars come and go, and is still actively involved in Winston Cup competition. Of all the race cars he's worked with, Hyde holds the Dodge Charger Daytona in the highest esteem.

"The strength of the Daytona, of course, was that wide aileron going down to hold the wing, which held it stable as far as going into the corners," Hyde points out. "Then you strung that wing across the back, and it was up high in the good open air. My goodness, it had good downforce, and then you had the big ol' piece coming down the side there that would hold the side forces in. You could hang that car out constantly two, two and a half feet and just keep riding and it would stay right there. It was a perfectly balanced car.

Bobby Isaac chases Jim Vandiver. Like most wing car drivers, Vandiver feels the 1970 Chryslers were far and away the best-handling stock cars ever built.

"We didn't try to change any design on that car. That windshield looked like it was bent but it was a perfect design for the nose slope," continues Hyde. "The air that came up over that hit that wing perfectly, up there where it was in the fresh air. It really got a hard charge of air."

On August 23 it was time for the Grand National cars to race at Talladega, which in 1970 meant it was time for Pete Hamilton to streak to victory again. Hamilton completed his sweep at Alabama International Motor Speedway in the number 40 SuperBird as he won the second annual Talladega 500 — this time run with no boycotts, Pogo Effects, or tire problems. Bobby Isaac, the fastest qualifier at 186.834 MPH, was second, while Charlie Glotzbach drove to third.

Pete Hamilton attributes his success with the SuperBird to the car's design, which moved the aerodynamic center of pressure rearward. "We could run softer springs in the front which allowed the tires to stick," says Hamilton. "If you ran softer springs the car was more gentle over the bumps and it kept the tire on the ground longer. If you keep the tire on the ground longer it sticks better and you can run faster.

"The pointed nose, all that did was reduce the amount of horsepower required to run a certain speed," Hamilton continues. "So if you reduce the amount of horsepower required, with the amount of horsepower we had available you're going to run faster down the straightaways. We didn't do much adjusting of the horizontal stabilizer. At speed, in a neutral position, one or two degrees of rake really made a big difference.

"What we tried to do at Petty Enterprises with this thing was to get the car to drive correctly and be tight enough without cranking that thing up. If you left it as an airfoil, and sprung the car properly, you could leave it flat and you had the least amount of drag for the most amount of downforce in a neutral position. And all of this was unheard of on a stock car."

The wing cars returned to action at Darlington Raceway for the famed Southern 500 on September 7. This time Buddy Baker found his way to victory lane in Cotton Owens's dark orange number 6 Daytona, winning a race his father, Buck, had won three times. Bobby Isaac had another consistent finish in his Daytona as he took second, ahead of Pete

Hamilton's Petty Enterprises SuperBird. Ford fared better in qualifying, with David Pearson's 150.555 MPH lap being good enough for the pole.

"You hear where drivers win races? Well, there's places where cars win, too," says Baker of his Daytona. "This wing car was the kind of race car that would win races for you. That was just awesome. Awesome is almost close enough to explain how good that race car drove. You couldn't lose it — you could not lose it. You'd just run as fast as your heart could stand it!"

For the wing cars it was on to the one-mile, high-banked Dover Downs International Speedway for the Mason-Dixon 300 on September 20. Richard Petty proved his dominance on medium-sized tracks with yet another win. Bobby Allison and Charlie Glotzbach were next in the top five, with the Fords of David Pearson and Benny Parsons rounding out the top positions. Bobby Isaac's number 71 Daytona was fastest qualifier with a speedy lap of 129.538 MPH.

LeeRoy Yarbrough came to the rescue of the Ford fans in his number 98 Junior Johnson Mercury when the National 500 was run at Charlotte Motor Speedway on October 11. Yarbrough beat Bobby Allison, who drove Mario Rossi's Daytona, and Fred Lorenzen, who drove the Ray Fox Daytona, to complete a race which ended under caution. Charlie Glotzbach upheld Chrysler's qualifying honor with a 157.273 MPH orbit of the 1.5-mile speedway.

On November 15, 1970, the wing cars competed against their Ford counterparts for the final time of the season. Even though Glotzbach again showed the way in qualifying with a lap of 136.498 MPH, the American 500 at North Carolina Motor Speedway in Rockingham was not one of the wing cars' shining moments. Cale Yarborough held off a challenge from David Pearson to win. Behind the one-two Fords were Bobby Allison in his Daytona and his brother Donnie in the Banjo Matthews Ford. Buddy Baker managed a fifth-place finish, followed by Richard Petty in his SuperBird and Bobby Isaac in his Daytona.

With that, the 1970 aero war came to a close.

The wing cars had been amazingly successful given the rushed debut of the Dodge Charger Daytona in September 1969. In 1969 and 1970, the Daytonas and SuperBirds had 14 wins on tracks of a mile or more in length compared to 10 for the Fords and Mercurys. The figures become even more one-sided when you look at the top-five finishes for the 1970 season. On tracks of a mile or more in length, the wing cars placed in the top-five finishing positions 61 times compared to 38 top-five finishes for Ford and Mercury. These figures include the two 1970 125-mile qualifying races for the Daytona 500 (Ford had one win and three top-five finishes compared to Chrysler's one victory and seven top-fives).

Even more outstanding than the wing cars' racing record is their qualifying record. Ford was the fastest qualifier at only three of the 1970 races on tracks over one mile in length, while the Chrysler cars started on the pole a commanding 15 times that season. The Charger Daytonas driven by Charlie Glotzbach and Bobby Isaac were the fastest qualifiers five times each. The wing cars were clearly the swiftest and strongest of the aero warriors.

The weekend following the Rockingham event, the final short track race of the season was held. When the points were tallied, Bobby Isaac had captured the 1970 NASCAR Grand National championship. Isaac's wins at the shorter speedways, combined with his numerous top-five finishes when the wing cars were deployed, earned him the title over Bobby Allison, James Hylton, Richard Petty, and Neil Castles.

Bobby Isaac was something of an enigma in the racing world. In an era of forceful personalities, Isaac was quiet and generally kept to himself. It was this quietness that had led some of the other Grand National drivers to believe that Isaac could not be trusted when the PDA was being formed in 1969.

"Bobby Isaac was a good friend of mine," says Jim Hunter, who was a newspaperman covering the Grand National beat during the wing car era. Hunter went on to become a vice president of NASCAR and president of Darlington Raceway.

"I remember a story I did on Bobby," Hunter continues. "When he was a kid he used to sit on a grassy hill near his home and watch the trailers with the race cars on the back of them come in to Hickory Speedway, and he always said, 'One of these days I'm going to drive a race car.' At an early age, he wanted to be a stock car driver."

Born on August 1, 1932, in rural North Carolina, Isaac was the youngest of nine children. "He grew up in hard times, very hard times," says Bill Brodrick, an Isaac associate who worked for Union 76 (now Unocal Corporation). "Racing was something he could do and be good at. Racing permitted him to use his talents in a job, whereas if he didn't have racing, he'd have been a mill worker. So he was very fortunate, and he understood that."

Like most racers, Isaac ran short tracks early in his career. "When I was a young reporter Bobby used to come to Columbia Speedway, which ran on an odd night — Thursday night — every week," Hunter recalls. "Bobby used to come down there and run the old flatheads against Dale Earnhardt's daddy, Ralph, and David Pearson, and Ned Jarrett."

As Isaac's driving skills flourished, he became a good friend of David Pearson's. "He was a sidekick of Pearson's," Hunter says. "They traveled together and ate together, that sort of thing. Bobby was quiet, and probably the biggest reason he was quiet was that he was uncomfortable with the spotlight due to his

There was no mistaking Chrysler's turf in the NASCAR garage area.

lack of formal education — which a lot of guys were back then. That's hard for people to imagine today, in this day of public relations and all that sort of stuff."

"He was a quiet type of guy," Brodrick affirms. "David Pearson was more outgoing than Isaac was, and Bobby fed off of David on that. I've been in a car with him and we'd go miles and miles and never say a word — just a very quiet guy. He was not shy, but he was self-conscious about himself and his ability to communicate and deal with upper-management type stuff. He wasn't comfortable in those situations. But if you had him by yourself or you were just working one-on-one, oh, he was a great guy."

Although Isaac was a very consistent racer on the big tracks, Hunter believes Isaac was never as comfortable there as he was on the short tracks. Hunter's opinion is borne out by Isaac's championship-winning statistics in 1970. Isaac won 11 times in 1970, all on short tracks, but he consistently finished in the top five in his Daytona on the superspeedways — and claimed five poles in the wing car.

"He would run up front and try to put his car in a position to be there at the finish," Hunter explains. "Bobby was not aggressive to the point that he tore up a lot of equipment, and he didn't take spectacular chances, but he was a charger."

Brodrick agrees that Isaac was a charger. "If a car could win, he'd be out there doing it. He didn't ask for any favors and he didn't get any. He was a real racer — he let his actions do all of his talking. The publicity, my personal opinion is that he didn't care one way or the other about it, but he deserved everything he got. He was a hell of a race car driver."

Bobby Isaac died on August 13, 1977. A racer to the end, Isaac was competing at Hickory Speedway in North Carolina. Running in third place — as always, in a position to win — Isaac signaled for a relief driver near the end of the race. Shortly after getting out of his race car, Bobby Isaac collapsed. He died later that night.

Brodrick refers to Isaac's relationship with crew chief Harry Hyde as a good marriage. "Harry and Krauskopf and Isaac were a good group that got together and permitted Bobby to exploit his talents," Broderick says. "Those people brought him out and they understood him. He and Harry were perfect together, absolutely perfect. Harry was like a father to Bobby, and Harry could talk to him."

Hyde also had another talent that helped the car number 71 team to the championship: "Harry Hyde seemed to listen the closest to what Larry Rathgeb had to say, and I think that showed," says Chrysler's Gary Romberg.

"That could have been," agrees Jim Hunter. "Back then, when you talked to most of the stock car teams about factory engineers, it was 'The concepts look good on paper, but do they work?' Harry listened to the engineers, and might have been more receptive to trying some of the things. Harry might have been ahead of his time. Harry had a way with his drivers, and still does. 'Let's try it. What have we got to lose? It might work.' Harry has always been an innovator."

Harry Hyde looks back with fondness on the wing car era.

"I've got some real inner feelings about that car. You just knew you had a real good car," Hyde says. "You knew you were the best when you pulled in. The feeling then was a lot different than it is now. We had the big ol' Hemi and Ford had the big ol' 429 and we knew it was going to be a hell of a race. We knew we had the shit to get them with and they knew they had the shit to get us with.

"The corporate world got into it, and I can't kick it — racing's been good to me," Hyde admits. "But don't you see, back then you were in an era where they built those big, high-banked tracks and they were coming up all over the country, and we were hitting them with a hell of a strong motor and a hell of a strong car. It was a feeling then that was different, and the Dodge was a very special part of us. That was a very special time."

Opposite: One of the greatest sights in NASCAR's history was the wing cars hurtling through a high-banked turn.

Wild in the Streets

When the decision was made to create the Dodge Charger Daytona as a 1969 model after the Charger 500 came up short in its Daytona 500 debut, Chrysler officials had unknowingly made their lives a lot easier.

In April 1969 the body that oversees auto racing in America — the Automobile Competition Committee of the United States (ACCUS) — met to raise the minimum production run required to qualify a car for NASCAR Grand National competition. Effective beginning with the 1970 season, a manufacturer wishing to qualify a car for NASCAR racing would have to build either 1,000 cars or an amount equal to one-half the number of franchised dealers — whichever was greater. By accelerating the Daytona program, Chrysler had qualified the car under the old rules, which would require a minimum production run of just 500 cars.

Still, there was plenty of work to be done to meet the corporate goal of having the winged car ready to race at Alabama International Motor Speedway in September 1969. While the engineers and aerodynamics people had their hands full designing the car and making it race-ready, those who had to ensure 500 street Daytonas were built in time to qualify the race car faced an even more daunting task.

In the automotive industry, the development of a new model usually takes years. With the Daytona program, Chrysler had only a few months to create a new, radical model — a model based on the design of yet another special car, the Dodge Charger 500.

Dodge product planner Dale Reeker, who had taken part in the initial Daytona program meeting called by Special Vehicles Group's Larry Rathgeb and George Wallace, was assigned the formidable task of overseeing the street Daytona program. Reeker had accompanied Rathgeb as they climbed the Chrysler chain of command, seeking approval for the winged Daytonas. They ended up with a "yes" from the vice president and general manager of Dodge, Bob McCurry. Reeker knew there was no margin for error in meeting the goals of this high-profile program.

Like other automobile manufacturers, Chrysler Corporation had occasionally turned to Creative Industries of Detroit for help with special projects. The firm specialized in automotive design, tooling, fabrication, and engineering services. Reeker had teamed up with Creative Industries to complete the Charger 500 project, and the firm looked to be the perfect choice to complete the Daytona program.

Opposite: If you saw this in your mirror, you knew you were in one of the greatest muscle cars ever built.

"Our largest customer was Chrysler Corporation, and we were doing design work and tooling and fabrication for Chrysler," says Verne Koppin, who founded Creative Industries with Rex Terry and Richard Leasia in 1953. (Koppin's father had been heavily involved with the legendary NASCAR exploits of the Hudson automobiles.) When the Daytona program arrived at Creative Industries, Koppin was the firm's chief engineer.

"It was a crash program right from the word go," Koppin points out. "When Chrysler assigned a vice president to take it and see that the schedule was met — I can tell you that puts the horsepower in the schedule! We had a deadline — the Talladega race — and these cars had to be built and approved by NASCAR. That had to be accomplished by a specific date. Believe me, it was a tight program and there were critical dates that had to be made or they wouldn't make that race."

Under the Daytona program's three-pronged approach with aerodynamic design, proving grounds testing, and street version manufacturing all occurring at once, Koppin and his company had their hands full. But the creation of the winged wonder added a brand new dimension to Creative Industries' work.

"The aerodynamics was done in the lab at Chrysler Corporation, and then they gave us that information," Koppin explains. "We did the clay modeling of the pointed front end, and we did all of the tooling in making the castings that went on the spoiler. We then fabricated the parts in our stamping

Right: Early advertising from a 1969 Alabama International Motor Speedway program. As the copy indicates, the Daytonas weren't yet available to the public.

Opposite: The design that made Chrysler king of the muscle cars.

The obvious reason Richard Petty came back.

SuperBird, the ultimate Road Runner.

It's not the only reason Richard returned to Plymouth, but you can put it at the top of the list. (We're building a limited number of completely *streetable* SuperBirds. Check your Plymouth Dealer.)

Another reason came from Richard himself: "The Pettys and Plymouth are like, well, like family." During an 11-year association with Plymouth, Richard won (among other things) two national championships, most total money and a legion of loyal fans. Richard's also won a record 101 victories, 92 of them in Plymouths.

Now the family's back together again. And we're glad.

The Pettys will prepare a couple of SuperBirds for the superspeedways and run the smaller Road Runners on the short tracks.

So we're officially back in NASCAR Grand National racing. With the most successful stock car driver in history.

Which is the sort of thing you could expect from Plymouth.

Because we have the most comprehensive high-performance program in the industry.

It's called the Rapid Transit System. And it offers everything from the cars themselves (Road Runner, GTX, 'Cuda, Duster 340 and Sport Fury GT), to factory tuning manuals and high-performance parts, all the way to TransAm racers, Super Stockers and AA/Fuel dragsters.

If you're a car enthusiast of one kind or another, you'll feel at home in the System. Because there's something for every kind of enthusiast.

It's sort of a family affair. In more ways than one.

Road Runner Character © Warner Bros.—Seven Arts, Inc.

```
CHRYSLER HGPK
TLXB DET
TLX171 BB151 B STA140 PD                      RECEIVED
FAX STAMFORD CONN14 1227P EST
                                           1970 JAN 14 PM 1 32
GALE W PORTER MGR ,HIGH PERFORMANCE CHRYSLER CO
HIGHLAND PARK GENERAL OFFICE BLDG          COMMUNICATIONS
                                           MESSAGE CENTER
  HIGHLAND PARK MICH

COMPLETION OF DEALER SURVEY BY ACCUSFIA COVERING THE PLYMOUTH
SUPERBIRD INDICATES THAT CHRYSLER   PLYMOUTH HAS COMPLIED WITH
THE PRODUCTION REQUIREMENTS   ESTABLISHED FOR 1970 STOCK CAR
RACING THE PLYMOUTH SUPERBIRD AS DELIVERED TO DEALERS IS
THEREFORE ELIGIBLE FOR COMPETION AS OF THIS DATE
```

DODGE ANNOUNCES SCAT CITY
The '70 Dodge Scat Pack is road ready.

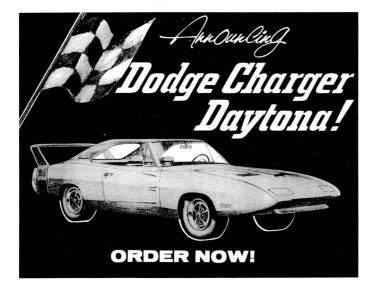

Preceding pages: The Petty Enterprises gang looks happy with this SuperBird, even if it is a street car disguised as one of Petty's Grand National racers. Dodge advertising was centered around a Bobby Isaac road test of the Daytona.

division, so we had all of the requirements to do everything but the assembly. Of course, if we'd be doing the clay modeling, if we'd be doing the tooling, if we'd be doing the fabrication, the logical thing would be to do the assembly of the vehicle. So we picked up this plant in East Detroit, and we set up the process there. We set up a small production line and went ahead and built the cars."

At a leased production plant at 17630 East Ten Mile Road in East Detroit, the Dodge Charger Daytonas grew their wings. Dale Reeker and the other Chrysler people and Verne Koppin and the rest of the Creative Industries team had to coordinate the complex tasks of the design, manufacture, and assembly. Among the challenges they faced were the nose cone, with its retractable headlights, a special scissors jack, side marker lights, and the distinctive vertical stabilizers and aluminum airfoil.

Before a single Charger Daytona rolled off the new production line, a press preview was held on April 13, 1969. Since there were no real Charger Daytonas yet in existence, one had to be made. A production Charger underwent a hurried cosmetic assault, eventually winding up with a fiber glass nose and spoiler assembly. There were no headlights in the car's nose — black tape outlined the light doors. Tape also replaced the rubber seals that would eventually be used around the nose, and a heavy dose of modeling clay was administered to fill a gap in the white car's hood.

In addition to "the car that was never real," as test engineer John Pointer refers to the hurriedly cre-

Right and opposite: Plymouth SuperBirds emerge from their Lynch Road nest.

ated white Daytona, Pointer's proving grounds test car was also shown off flying the colors of the K&K Insurance number 71 to approximate what the world would hopefully see in Talladega, Alabama, come September.

The press went wild over the radical Daytona, practically killing themselves to come up with new adjectives to describe the winged wonder. But more importantly, dealer orders for the Daytona began to flood in — even though Chrysler lost money on each car sold as a result of the special parts required for every Daytona built.

"We had 1,200 orders in three weeks for 500 cars," says John Pointer. "The pricing of the Daytona was such that it was only a few hundred dollars over that of the standard Charger R/T, so the dealers could be sure to sell them. The euphemism was that there was a 'net reduction of variable profit' of $1,500 for each Daytona that was sold. You were talking about a special trunk lid, special back light, the nose cone, modifying the front fenders with the scoops . . ."

The Daytona package was an expensive one to produce, and at a loss of $1,500 per car, it was clear that profit wasn't Chrysler's key consideration. Making the Talladega race was, and the cars had to be delivered to dealers by September 1, 1969. At Creative Industries' production plant, Don Mirzaian was in charge of making sure there were no delays.

"Don was the man on the floor, and he saw to it that the job got done," Verne Koppin says. "We'd have periodic meetings a couple of times a week to

Under the hoods of the winged cars lurked a variety of muscular motors. Clockwise from top left: Plymouth 440 4-bbl., Dodge 440 4-bbl., Plymouth 426 Hemi, and Plymouth 440 Six-Pack.

verify schedules and identify any problems. Mirzaian was the whip to see that those cars were pumped off that production line and going out there."

"When we were doing the Charger Daytona, we started out doing a couple a day," recalls Mirzaian. "Eventually, we got up to about 20 a day. We got them all out in less than three months. They had to have them all out or they couldn't race."

Several hundred 1969 Dodge Charger R/Ts made the trip from Chrysler's Hamtramck plant to the leased Creative Industries facility to make the metamorphosis into winged Daytonas. Exterior paint choices and interior trim colors were the same as those available on the non-winged Chargers, but air conditioning was not available because at low speeds it would have created a severe engine cooling problem. Daytonas were built with both four-speed manual transmissions and TorqueFlite automatics, and, with the exception of 70 426-cubic-inch Hemi-powered wing cars, the powerplant was the massive 440- cubic-inch Magnum. The base price for a Dodge Charger Daytona was $3,993 — a true bargain.

Most sources agree that, by the end of the project, a total of 503 Dodge Charger Daytonas were constructed. Verne Koppin notes, "NASCAR came over and counted them to make sure we had a physical car to meet their minimum number."

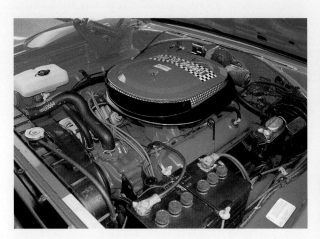 

Creative Industries also had to manufacture the parts kits used to convert the racing Charger 500s into Daytonas. Dale Reeker would then supply the parts to Dodge performance coordinator Bob McDaniel, who was responsible for distributing them to the Dodge racing teams.

The SuperBird project hatched a whole new nest of problems when Plymouth began planning to race it in the 1970 NASCAR season. The new ACCUS rules governing 1970 required that more than 1,500 winged Plymouths be constructed.

Due to the high volume of cars that had to be built, the parts for the SuperBird program would still be made by Creative Industries, but the actual car construction would take place at Chrysler's Lynch Road assembly plant. At a September 24, 1969, meet-

This ad features the car displayed at the Daytona's media preview. Among other illusions, the headlight doors are nothing more than black tape.

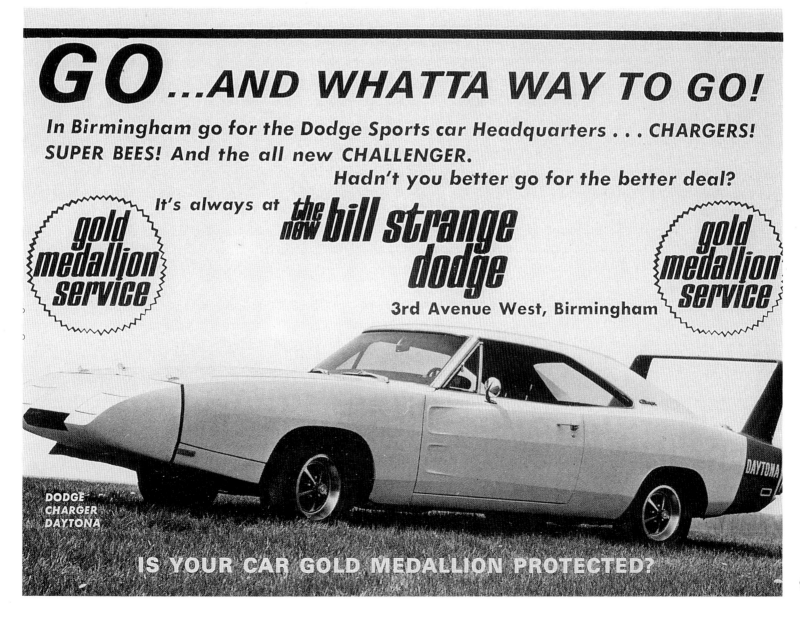

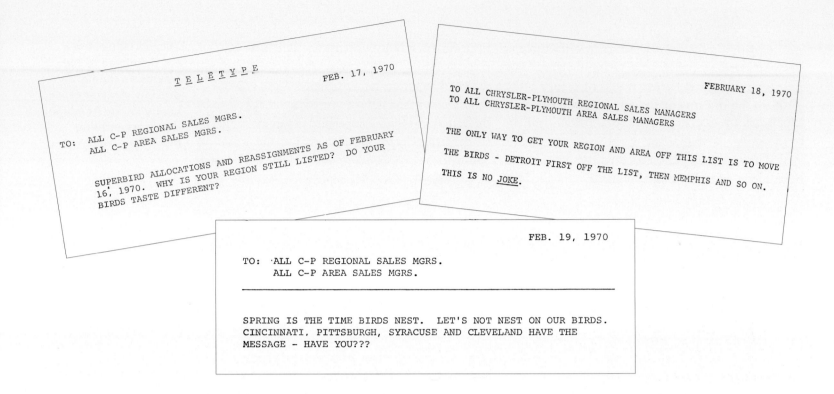

Above and opposite: These memos reveal the corporate distress over the sluggish sales of SuperBirds. Meanwhile, the winged inventory gathered ice awaiting owners.

ing of the coordinators of the "Belvedere-Daytona Program" — as the SuperBird was initially known — the Sales Programming staff made plans to sell 1,920 SuperBirds. Production would begin with 10 units in October, followed by 1,160 in November, and would taper off to 750 in December.

Plymouth had hoped that much of the Daytona work could simply be grafted onto the Road Runner, but that was not the case. John Herlitz and the Plymouth styling people pointed out that a clash of body lines would result from any hurried attempt to put a Daytona nose on a Road Runner. Because there would be such a large production run of SuperBirds, more attention had to be paid to styling and marketing considerations than had been necessary for the Daytona program.

In the end, Dodge's Coronet front fenders were added to the SuperBird. The Coronet also contributed a heavily reworked hood to the SuperBird effort. The vertical stabilizers on the SuperBird received a 40 percent size increase and featured a more dramatic rake. John Herlitz came up with a unique finishing touch.

"One of the cute sidebars of the SuperBird story was the big Road Runner circular decal on the tail fin," Herlitz says. "In the shop out at Creative Industries on the afternoon we were trying to put all of that together, I couldn't find a compass to draw the circle on the vehicle — so I ended up tracing that size and that circle off a wastebasket!"

The SuperBirds, unlike the Daytonas, were constructed from scratch at Lynch Road using the parts Creative Industries built.

"On the SuperBird we did everything in primer and sent it down to the pilot plant," recalls Don Mirzaian. "We made all of the parts, the nose cones with the headlamp assemblies in them, and all of the pieces. Then they did the final painting."

The painting process yielded a nasty surprise for general manager Harlan McDonell and the workers at the Lynch Road plant during one phase of the SuperBird project. "One of Chrysler's vendors had sprayed on a non-rusting compound on the parts," recalls Don Mirzaian. "They sprayed it on without ever letting us know. When they did the final paint-

TO: ALL CHRYSLER-PLYMOUTH REGIONAL SALES MANAGERS
    ALL CHRYSLER-PLYMOUTH AREA SALES MANAGERS

HAPPINESS IS A SUPERBIRD WINNER. PETE HAMILTON IS HAPPY - YOU CAN BE ALSO.

SELL THE BIRDS.

FEBRUARY 20, 1970

TO: ALL CHRYSLER-PLYMOUTH REGIONAL SALES MANAGERS
    ALL CHRYSLER-PLYMOUTH AREA SALES MANAGERS

CONGRATULATIONS - WHERE THERE IS MOVEMENT, THERE IS LIFE. CHICAGO WIGGLED, CAN YOU?

TELETYPE

MARCH 2, 1970

TO: ALL CHRYSLER-PLYMOUTH REGIONAL SALES MANAGERS
    ALL CHRYSLER-PLYMOUTH AREA SALES MANAGERS

SINCE R. K. BROWN'S SUPERBIRD WIRE OF FEBRUARY 18TH, 41% OF THE REGIONS LISTED HAVE FAILED TO MOVE A BIRD. DO YOU REGIONS SOMEHOW CONSIDER THIS A JOKE?

REGION	ALLOCATION	R/A	% OF ACCOMP.
MEMPHIS	13	10	76.92
BOSTON	14	8	57.14
CHICAGO	31	16	51.61
PORTLAND	6	3	50.00
JACKSONVILLE	11	5	45.45
NEW YORK	20	9	45.00
LOS ANGELES	24	10	41.66
WASHINGTON	15	6	40.00
PHILADELPHIA	15	6	40.00
ST. LOUIS	8	3	37.50
DENVER	6	2	33.33
ATLANTA	22	7	31.81
SAN FRANCISCO	10	3	30.00
MINNEAPOLIS	4	1	25.00
KANSAS CITY	25	3	12.00
DALLAS	20	2	10.00

**It's a bird!
It's a car!
It's the Road Runner!
Superbird**
from Plymouth's
Rapid Transit System!

Ordering Data:
Mechanically, the Superbird is virtually identical to standard Road Runner hardtops, except that a "GTX"-type 440 V-8 with corresponding driveline and suspension is standard equipment. Normal ordering procedures should be followed, except where noted.
The following equipment is standard on the Road Runner Superbird. However, it must be coded on your order sheet. These codes are as follows:
- Sales Code (A131)
- Black Vinyl Roof (V1X)
- Competition-type Hood (J45)
- Performance Axle Package (A36—Automatic), or Track Pak (A33—Four-speed)
- Power Disc Brakes (B41 and B51)
- Power Steering (S77)

Special Note:
The following options are *not* available on Superbirds:
- Rear-seat Speaker (R31)
- Air Conditioning (H51)
- Air Grabber (N96)
- Sill Moldings (M25)
- Light Package (A01)
- Rear Window Defogger (H31)
- Performance Hood Stripes (V21)
- Headlight Delay (L42)
- Basic Group (A04)
- Sure-Grip Differential (D91)
- Axle Packages (A31, A32, A34)
- Trailer-towing (A35)

In addition, seat availability is limited to black or white bench or bucket seats.

This brochure (above) was designed to entice dealers to order SuperBirds, while another ad (opposite) was obviously aimed at a different audience.

ing, all of the paint started peeling! We had to sand them all down and re-do them, and still get them all out in three months."

At the peak of the SuperBird manufacturing process, more than 50 cars per day were being completed to meet Plymouth's January 1 deadline — the date that a new, restrictive government headlight ruling was to take effect. The winged Plymouths were coming off the line at double the pace of the Daytona program.

All of the SuperBirds featured a vinyl roof to hide the scars of surgery that gave the car a sleeker backlight than its non-winged relatives. The winged Plymouth was available in eight colors, with snazzy names like Tor-Red and Lemon Twist. To speed up the manufacturing process, some minor vehicle options that were available on the Road Runner, such as a rear defogger, hood performance paint, and headlight delay, could not be ordered on the SuperBird. Still, there was an impressive choice of engines available. The 375-horsepower 440-cubic-inch motor was standard, the Six-Pack 440 and 426 Hemi were optional. As with the Daytona, air conditioning was not available on the SuperBird.

The most commonly agreed upon production figure for the Plymouth SuperBird is 1,935 units built. Despite the confusion over the official total, there's no question there were many more winged Plymouths than Dodges. In 1970, that created a problem at many Plymouth dealerships.

"Plymouth found out there were about 500 people willing to buy these things and they had all already bought the Dodges!" jokes Special Vehicles Group's George Wallace. "They had a lot of trouble getting rid of the SuperBirds. Some of the Plymouths were converted back to standard Road Runners with a stock front end, but they still had the wing. The dealers got stuck with them and they did everything they could to sell them."

Memos from corporate headquarters to Chrysler-Plymouth Regional Sales Managers contain warnings such as, "Since February 18th 41% of the regions listed have failed to move a 'Bird. Do you regions somehow consider this a joke?" and, "Spring is the time birds nest — let's not nest on our 'Birds." When Pete Hamilton won the Daytona 500 in a Petty Enterprises wing car, the corporate word went out quickly: "Happiness is a

SuperBird winner. Pete Hamilton is happy — you can be also. Sell the 'Birds!"

But sluggish sales weren't the only problem confronting the SuperBird — several states had a problem with the car's design. Maryland, for example, refused to register the cars or issue certificates of title because a state law required that all vehicles have front bumpers. In addition to prohibiting the special Plymouths on Maryland highways, the few dealers who had already sold SuperBirds in Maryland were urged to contact the owners and make arrangements to have the cars modified to comply with the state law. It's unknown if any owners surrendered their SuperBirds to do so.

Regardless of sales figures and legal restrictions, the wing cars turned heads wherever they went. By 1970 the public had grown accustomed to muscle-bound oddities, but the Dodge Charger Daytonas and Plymouth SuperBirds set new standards for the muscle car era with their extreme styling and highly potent engine combinations.

Sales literature for the Daytona centered around Grand National driver Bobby Isaac taking the car for a test run at the five-mile Chelsea Proving Grounds track.

"Well, there's one obvious thing about a Charger Daytona," the ad begins. "Nobody, but nobody walks by without breaking his neck to take a second look. This is the slightly civilized version of the shark-nose, built specifically for the long NASCAR ovals. Old Slippery has a snout that strikes out a country mile in front, and an adjustable spoiler that looks two stories tall in the rear . . . a car that you'll never lose in a crowded parking lot."

The SuperBird sales pitches also stressed the NASCAR connection, and naturally enough, Richard Petty was at the center of the campaign. A two-page, black and white ad that showed a tough-looking Petty Enterprises crew gathered around the number 43 wing car read: "The family's back together again, and we're glad. The Pettys will prepare a couple of SuperBirds for the superspeedways and run the smaller Road Runners on the short tracks. So we're officially back in NASCAR Grand National racing. With the most successful stock car driver in history."

Plymouth also ran a rather bizarre ad that told the tale of one "Lightning Billy," who used his SuperBird to run moonshine. Illustrated with a photo of a winged Plymouth parked next to a still, the ad boasted: "Lightning Billy's business coupe, with an enormous wing and a long, pointed snout, and great fiery eyes, moaning like a banshee as it sheds the feds in the hills."

Advertising claims aside, public relations director Frank Wylie notes that the wing cars were a very distinct part of history. "There were several things that were involved, one being a new sense of aerodynamics," says Wylie.

"You know how if you drive down the highway and you go by a big truck you get that push-and-pull feeling?" Wylie asks. "In the Daytona, you'd just whip by those and you didn't even know you were going by. It was that good and that safe. It seems to me that a manufacturer has two basic responsibilities — one, to provide a safe car and two, to provide a fast car. There had been aerodynamic stuff on Ford's part and Chevy's part and so on, but at that point we just jumped ahead a generation. It was a lot of fun, and it was a great car."

For Verne Koppin, the wing car programs helped establish Creative Industries of Detroit as a force in the limited production automotive market — work carried on today under the name MascoTech Automotive Systems Group.

"That was a tremendous program," Koppin reflects. "That probably influenced our going into limited productions as a company. This was the catalyst that kicked things off, and it showed we could get involved in this limited type of production for component assemblies and so forth, if not complete cars. It was a tremendous program and an interesting program."

The effect that the wing cars had on the public in general, including a young racer who would

one day go on to win the Daytona 500, was also interesting.

"In 1970 Utica Club Beer sponsored a lot of modified races around New York state, and they used a SuperBird as the pace car for all the races they sponsored," remembers Geoff Bodine. "If you won a race, your name went into a hat and at the end of the year they were going to draw the winning driver who would get the car, and I won. I only drove the car in the summer for a long time, trying to save it. I loved the car. I mean, it was great! That thing would fly."

As exotic as the wing cars were, they weren't immune from the everyday dangers any automobile faces. Geoff Bodine's SuperBird was no exception.

"It was sitting in my driveway when I was living with my parents right after I got it," Geoff recalls. "My father walked out one morning, jumped in his station wagon — he didn't notice my SuperBird was sitting behind him. He backed right into the nose!"

SuperBirds awaiting shipment to Plymouth dealerships.

Where No Man Has Gone Before

It has a magical sound to it: 200 MPH. It conjures up visions of almost unbelievable speed. Sure, 199.466 MPH is plenty fast. In fact, 199.466 MPH was good enough to put Charlie Glotzbach on the pole for the Charger Daytona's debut at the first Talladega race. But 200 MPH? Now that's really fast. That's magic.

As the 1970 aero wars intensified, the minds of the NASCAR Grand National teams settled on the race-to-race battle for championship points. But the mind of Frank Wylie, Dodge's public relations director, was focused on that magic number: 200 MPH.

Despite the intensity of the speedway battles in early 1970, the official 200 MPH record was still waiting to be broken. Although the wing cars had frequently surpassed that mark in tests on the five-mile Chrysler Proving Grounds track in Chelsea, Michigan, those tests didn't count. In the automotive world, if you want to set a record you have to make sure the right people are there to document it. Without the proper verification, you could run over 200 MPH all day and it would be like it never happened. Officially, no one had ever run a 200 MPH lap on a closed course in any kind of automobile.

Frank Wylie, whose job was to keep the Dodge name in the public eye, decided to do something about this historic record that was just waiting to be broken. In February 1970, Wylie phoned Special Vehicles Group's Larry Rathgeb with a question: Could the engineering Charger Daytona break the 200 MPH mark at the track in Talladega?

Larry Rathgeb knew he had the weapon that could blast past 200 MPH. After all, the engineering Daytona was the same car that Glotzbach had used to flirt with 200 MPH at that first Talladega qualifying session. Rathgeb told Wylie that he was sure they could break the barrier and make the record official, once and for all.

Wylie's duties at Dodge included the assignment of Grand National drivers to the Dodge factory teams, so he would be the one to pick the man who would pilot the Daytona on the record run. Wylie knew there was one driver who wouldn't

Opposite: Buddy Baker prepares to make history on March 24, 1970.

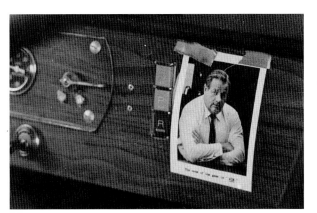

Baker drove under the stern gaze of Dodge boss Bob McCurry, whose photo was taped to the dashboard of the engineering Daytona with the admonition, "The name of the game is WIN."

NASCAR founder Bill France jokes with Buddy Baker (above), but the official NASCAR timing equipment (opposite) was no joke.

think twice about running hard enough to get past 200 MPH. His choice?

"I knew Buddy Baker was the guy to go for it," Wylie says.

"Baker, he loved to go fast," Larry Rathgeb elaborates. "The faster he went the better he liked it, and he liked our cars. The engineering cars were usually smoother than some of the others, than some of the race machines. When you wanted Baker to go really fast, when you thought everything was right, you'd say, 'Hey Baker — let's qualify this car. Do you think you can qualify this car?' He'd say, 'Yeah! Yeah!'"

Sure, he could qualify the car for a race, but could Buddy Baker qualify the engineering Daytona for the history books?

It is March 1970. Cold, wet weather is causing NASCAR some headaches; rain has postponed the scheduled running of the Grand National race at Atlanta International Raceway. But curiously, instead of heading for home, NASCAR chief inspector Bill Gazaway and chief timer and scorer Joe Epton travel to Alabama International Motor Speedway on March 24. One might wonder why two of NASCAR's top officials show such interest in a scheduled Chrysler transmission test at the 2.66-mile superspeedway. The answer is that Chrysler's use of the Talladega track is a "transmission test" in name only. Larry Rathgeb, Buddy Baker, and the Chrysler team are going to try to break the 200 MPH barrier.

As Epton sets up his timing equipment, Baker and Rathgeb look over the Daytona. Both men are intimately familiar with the sky-blue car — this is the vehicle in which all of the final innovations that the wing car program had brought to racing were tested. It seems only fitting that number 88 should have the chance to make motorsports history with the first official lap on a closed course at over 200 MPH. It also seems fitting that Baker, one of the two chief test drivers for the wing program, has been picked to guide the engineering car around the high banks.

It had rained the day before and that morning, but as noon approaches on March 24 the sun appears and the track begins to dry. Buddy Baker lowers his six-foot frame into the Daytona and brings the engine to life. The wing car rumbles down pit road and onto the track.

Baker cruises around the track four times before bringing the Daytona up to speed. His fifth lap is 191.985 MPH, and before bringing the car into the pits for adjustments on the eighth lap, Buddy is up to 194.200 MPH. Larry Rathgeb's notes of Baker's comments from the first run read, "Car feels good — it's flat down the straights. Wide open all the way. Tends to dart just a hair going into one and coming out of two and four."

The crew adjusts the suspension and carburetor jetting, and Baker roars back on the track. After running 197.839 MPH on the 12th lap of the session, Buddy pulls in again and reports to Rathgeb that the "Car drove a lot better. No darting now at all."

In the third session, the speeds climb to 198.322 MPH. Baker reports that "The car feels damn good."

In front of the empty grandstands of Talladega's super-speedway, Buddy Baker races towards the 200 MPH record.

In between runs, the engineers have handed Baker a roll of duct tape and instructed him to tape the car anywhere he thinks it might help improve the speed. The Daytona's camber is reset to make the car handle better, and Baker makes an early afternoon run at 198.850 MPH. The only problem is "a slight push off two."

With the wing adjusted, new tires in place, three inches of additional covering on the grille to cut wind resistance, and the Hemi's timing slightly changed, Buddy Baker makes the fifth run of the day at 3:20 p.m. Lap 26 is tantalizingly close to the magic mark — 199.085 MPH. Baker comes in, and Rathgeb's notes read, "Car moves better than before, push coming out of two was gone."

Rathgeb and the crew fine-tune the car further. "Add camber to left front wheel, remove camber from right front wheel. Toe set to 1/8" out, add one degree to timing, add 5 PSI to rear tires." At 4:25 p.m., on the 28th lap of the day Buddy Baker heads back out on the track to begin the sixth run.

The massive Hemi engine begins to build power as its RPM climb towards 6,500. Baker sweeps out of turn four and roars through the tri-oval towards the start/finish line to begin lap 29. The blue Daytona charges through turns one and two, down the long back straightaway, and heads for the end of the lap: 199.879 MPH.

Buddy Baker begins the 30th lap and runs a perfect line through turns one and two — right past Frank

Buddy Baker prepares for another run in the Chrysler race engineering Daytona.

Wylie and his camera on the outside of the turn. "I remember lying up there on the outside of the track, and then the thing went by," says Wylie. "It moved a lot of air, I'll tell you that. It was pretty exciting."

Baker runs straight and true down the 4,000-foot backstretch, and sets the car beautifully to enter the third turn. Through three and into four, the Daytona slices through the air as it whips around the 33-degree banking of the track. Out of turn four, Baker closes on the timing equipment — and history. Lap thirty: 200.096 MPH. Buddy Baker in the Charger Daytona becomes the first driver to officially break the 200 MPH barrier.

"We knew we could get the record," Frank Wylie says. "It was just a matter of Buddy taking a few laps and getting used to it and then nailing it. Buddy was the one that was doing it, though. I was just watching it and photographing it. He's the one that made it happen."

Chrysler's alleged transmission test resulted in a new world record. "We wanted to be the first race car to really take an official time over 200 MPH, of all kinds of cars," Buddy Baker admits. "I just went out and shook the car down, and I mean it was right on the money at 200. That doesn't really sound big until you think about what year it was. When you think about how we were on an eight-inch tire when they run 10 now, that the car was legal height and NASCAR-legal ... we came around that time and I went 'Whoa!' Quite frankly, it's just as well I didn't know when I did it!

Special Vehicles Group's Larry Rathgeb confers with Buddy Baker. Moments later, Baker shattered the 200 MPH barrier.

The record-breaking team. Left to right: George Wallace, Fred Schrandt, Buddy Baker, Larry Rathgeb, Larry Knowlton, and Gary Congdon.

"I've held the record at Daytona for years now for the fastest race time when I won the Daytona 500, but that will someday go down," Baker reflects. "That's why they call it a record — you can break a record. But nobody will ever beat me at 200 MPH. A milestone in the history of the sport, they can't take that away from you. It was a good race car and a good group of people, but it didn't really mean that much at the time. But now, as I get older in my sport, all of a sudden it means as much as anything I've ever done in racing. It's just as important as winning four times at Talladega like I have, or four times at Charlotte, or whatever — simply because of the unique car and the situation."

Naturally, being racers, thoughts turned to driving 300 MPH immediately after Baker went 200 MPH. "They said, 'Hey — the next barrier is 300. You want to do that?'" Baker laughs. "And I said, 'Hell no! No, thank you!' The next one's 300, and I want to meet the guy who does it, because he's going to have a pointed head, I believe!"

Just after 5:00 p.m. on that same day, Buddy Baker, who had spent much of the day diligently taping the Daytona to make it sleeker, hit the fastest speed of the session on the 34th lap: 200.44795 MPH. Car 88 had been stripped of all of Baker's careful taping right before this seventh and final run of the day when Larry Rathgeb decided to see what speeds the car was capable of with no tape. When Baker drove back in the pits after three consecutive laps over 200 MPH, he climbed from the car and dumped his roll of tape into a trash can.

Ironically, the next day was spent trying to make the Charger Daytona go slower. NASCAR hoped to make Grand National stock cars race at safer speeds, and one of the methods being considered to achieve this was to remove the cars' right side windows. Baker and the Daytona, after a couple of runs near 199 MPH, spent the rest of the session running with no side windows, about 3 MPH slower than the record-breaking pace of the day before.

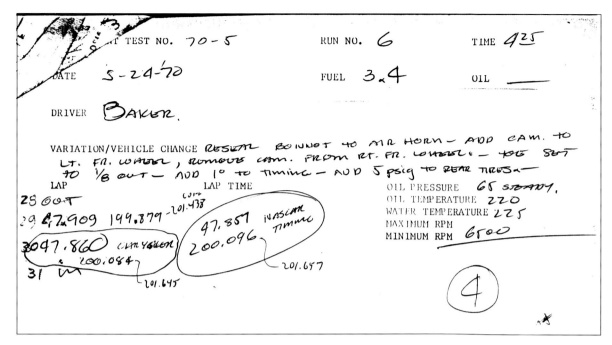

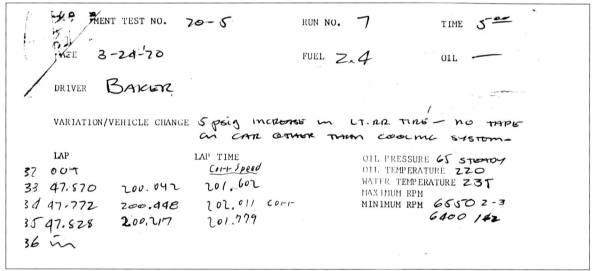

Larry Rathgeb's data sheets document the Chrysler team's historic speed assault. Lap 30 was the first run past 200 MPH, while lap 34 was the fastest of the two days. The higher "corr speeds" were calculated by Rathgeb to show how fast the car would have run under neutral conditions.

Buddy Baker has many fond memories of the engineering wing car. "The motor was built by Chrysler, the car was set up by Chrysler, and we went down to do the test. From the time we unloaded the car we knew we had something special going. I mean it — this thing was awesome!"

When the number 88 Dodge Charger Daytona left Alabama International Motor Speedway on March 25, it did so as the first car to break 200 MPH on a closed course. What most people probably don't know is that this most famous of all wing cars very nearly spent its existence rusting on the scrap heap.

"The car that was the Daytona, it's an interesting story of where the car came from," reveals Frank Wylie. The Dodge team needed a car body for the

race engineering Daytona when the development program got underway.

"We had a Dodge Charger 500 Hemi that was in California," continues Wylie. "It had been appropriated one night when it was out on loan to a magazine. It was finally found down in Watts on four milk crates, minus everything. That was about the time of the Riverside race, so one of the guys had brought a car out to sell and he had an empty trailer going back. So we put it on that trailer and took it back. That car eventually became that Dodge Daytona."

Saved from an ignoble fate and transformed into the first real racing Daytona, number 88 now resides in the International Motorsports Hall of Fame and Museum, appropriately located at Talladega Superspeedway in Alabama. Right next to the engineering Daytona is another famous record breaker — the number 71 K&K Insurance Daytona driven by Bobby Isaac.

At the end of 1970, Bobby Isaac wanted something more than the Grand National championship he had just won — he wanted to break Buddy Baker's speed record. On Tuesday, November 24, 1970, Nord Krauskopf's K&K Insurance race team pulled into Alabama International Motor Speedway to take their own shot at the record books.

"Bobby Isaac really wanted to do that," recalls Harry Hyde, legendary NASCAR crew chief and the head of the Isaac team. "He knew he could drive Talladega faster than anybody. We had won the championship, and he asked to go down there. Nord told him to take it down there and we'd run it. It was just a favor to Bobby."

The Daytona was unloaded in howling wind beneath the cold blue Alabama sky. The temperature was hardly ideal for setting speed records — the thermometer registered a bone-chilling 18 degrees — but Isaac had come to set a record and was determined to do it.

The first laps, like Buddy Baker's eight months earlier, were in the low-190s. Isaac picked up the pace to 195 MPH, but was having trouble with the stiff wind blowing around the 33-degree banking. Hyde changed the sway bar and suspension setup, and the next laps brushed past 199 MPH. More minor adjustments to the engine and suspension followed, and a new set of tires was mounted on the wing car. Isaac returned to the high banks.

On the second lap after Hyde's latest round of tuning the Daytona flashed by at over 200 MPH, its speed climbing higher despite the cold and wind. When he finished the twenty-second lap of the day, Bobby Isaac held a new world record — 201.104 MPH. An elated Isaac climbed from the car and was greeted by a chilled, but thrilled, crew.

"We went down there and we would probably have run 210 or 215 MPH that day the way we were fixed up," Hyde speculates. "But it was the coldest, windiest day I ever saw in my life! We absolutely could not get that car warmed up to run. It was so cold, it was pitiful. We were tickled just to break that record and come on home, because we sure did pick a bad time!"

Now that the K&K team had developed a taste for world records, they took the next logical step — the number 71 Daytona was hauled to the legendary Bonneville salt flats in Utah in September 1971. Harry Hyde had set his sights on the stock-bodied car world land speed records sanctioned by the United States Auto Club, and Bobby Isaac and owner Nord Krauskopf were more than happy to go along with the idea.

Opposite: Buddy Baker proudly displays the fastest speed of the Talladega record run (top). Thanks to some ingenious modifications (bottom), the truck that carried the engineering Daytona was capable of launching clay targets for skeet shooting — a concession to Larry Rathgeb's favorite hobby.

Low and lean, the K&K Insurance Daytona stalks records with Bobby Isaac at the wheel.

"Nord asked me, 'What do you want to do?'" remembers Hyde. "I said 'Let's go to Bonneville and set some records while we've got something to do it with.' So he backed the deal and a couple of months later I was on my way to Bonneville."

The salt flats of Bonneville are known for speed, and Isaac and his winged warrior fit right in. The desolate location fit Isaac to a tee, for with no distractions the Grand National champion could focus on just one thing — going fast.

Bill Brodrick, who attended the record runs as a representative of Union 76 Oil, says, "Bobby was uncomfortable with the press and didn't really like to get out and work hard doing publicity. That's one reason why I think he liked Bonneville — because there wasn't anybody there! It was a lot of sitting around and waiting because they'd want to get the salt right, and they'd want to get the timing right."

When attempting the speed record runs at Bonneville, the car did not have to conform to guidelines as stringent as those of NASCAR. Harry Hyde and his crew had more leeway as far as the Daytona's setup was concerned, but the wing car was so strong that little was required in the way of changes.

"We set that car up absolutely stock," recalls Hyde. "We had a 426 engine. We did not go with the big engine. I think we were allowed a 480 or something, but we used the 426 just like we raced it. The only thing we did was we lowered the car down probably an inch lower than you were allowed to run in NASCAR, but that's about the only thing. We didn't cheat it — it would have almost run through NASCAR inspection."

George Wallace of the Chrysler Special Vehicles Group had a history of going along for observation rides during Charger Daytona and SuperBird testing, so it shouldn't have been a surprise that he joined Isaac in the cockpit as the number 71 car was being set up. To Bill Brodrick, however, it was a big surprise.

"They had some problem and George wasn't happy with it," Brodrick recalls. "I'll never forget this because he got in the damn car and he rode with

Opposite and above: Scenes from Bobby Isaac's speed record runs at the Bonneville salt flats. Isaac broke 28 records in four days.

Opposite: The record breakers as they appear today, Buddy Baker's number 88 (top) and Bobby Isaac's number 71 (bottom).

Isaac! He had his feet up, wrapped around the bars — they had instruments in the car and he wanted to check those instruments. He hung on and they took off and I said, 'That man is out of his mind!' They ran that circuit with George hanging on reading pressure gauges or whatever to see what was going on. Then George got out and he was like, 'Oh well.' And I'll never forget that as long as I live — he got out there and ran 200 MPH in that thing!"

The low-key Wallace has particularly vivid memories of preparations to run on the 10-mile Bonneville course.

"The condition of the salt that year was such that rather than the 10-mile circle that is the preferred track at Bonneville, they had to run a 10-mile oval," Wallace says. "They were basically two-mile straightaways and three-mile turns. I rode with Bobby while we were setting up the car a little, and he was probably the ideal driver because he was a dirt tracker who wasn't afraid to go fast. He was basically driving it like a huge dirt track.

"He'd get up to about 205 or 206 MPH at the end of the straightaway and he'd never lift. He'd throw it into the turn and from the inside it felt like it was going out about 30 percent. The tail end would hang out, but he would drive it just like you would on a dirt track. At first I was a little anxious about it, because we had to point out to him that if you lose the car you could spin for five miles."

"Isaac had some great stories from the salt flats," remembers Jim Hunter, a friend of Isaac's. "He'd sort of laugh in his dry way and say, 'You didn't even know how fast you were going. You'd about go to sleep because you were going in a straight line for so long!' He said you got real comfortable because there was nothing to hit."

There may have been nothing to hit, but as Bill Brodrick notes there was still more than enough danger. "It was a serious thing if you got in trouble and got off the salt and got crossed-up where they didn't grade it. All they did was mark it and grade the inner circle on the salt. If you got off where you were running and out of the groove, you could get killed real quick. It was serious running that fast," Brodrick emphasizes.

Indeed, Bobby Isaac did have one very close call. While running the 10-mile oval, Isaac lost control at over 200 MPH and launched the Daytona over a 10-foot ditch that had once spelled disaster for jet car driver Craig Breedlove. Breedlove's famous "Spirit of America" had been destroyed, but Isaac was more fortunate — the K&K Dodge landed with little damage.

The work of breaking land speed records got underway for real on September 13, 1971, when USAC land speed record chief steward Joe Petrali and his crew began manning the timing equipment. Bill Brodrick found the spectacle of Bonneville to be far different from the Grand National racetracks.

"The car would go out of sight, and you could hear it way, way over the mountain range," Bill remembers. "Then you'd see this little speck like a gnat. We were at the far end, in one little group, and when Isaac would come down he'd be coming by you but he'd be looking straight at you! It looked like he'd be crossed-up but he'd be making that gradual turn. I don't think Bobby ever really got out of shape, but he would throw up a rooster tail. We were all standing inside and here comes Isaac — zoom! He was on the ragged edge all the time because of all that horsepower he had, but of course Bobby was one of the greatest dirt track drivers there ever was. Isaac ran that thing loose as a goose."

Was Isaac's dirt-trackin' heritage the secret to conquering the wide-open spaces of Bonneville? The answer can be found in the USAC record books. The K&K Insurance Daytona left shattered speed marks in its wake as the winged car thundered across the salt. In all, 28 new speed records were set during the four-day assault on the salt. Among the highlights were a one-mile pace of 216.946 MPH, a breathtaking kilometer speed of 217.368 MPH, a 100-mile run at 194.290 MPH, and a standing-

A detailed look at the interiors of the record-breaking stock cars of Buddy Baker (blue) and Bobby Isaac (orange).

start, 10-mile blast of 182.174 MPH. Many of the records Bobby Isaac set at Bonneville still stand today.

USAC's Joe Petrali, noting the poor condition of the salt, said, "I have never seen any driver dirt-track around the 10-mile oval before."

To which Isaac replied, "Maybe I ought to get credit for going 200 MPH sideways!"

George Wallace credits one of the Harry Hyde team's very best motors for the mind-boggling number of speed records that were obliterated in such a short time.

"USAC at the time allowed the Hemi to run two four-barrel carbs, where NASCAR limited us to one," Wallace notes. "The engine in the car was the one Harry Hyde used to qualify at Daytona and Talladega. You can build two identical engines and there's only one good one — you could measure every piece on it and you'd never know why. But that was Harry's good qualifying engine, and it did good for us at Bonneville like it did everywhere else." Bobby Isaac summed up the experience simply. "To be here is one of the greatest honors I've had," he said.

"The secret to him breaking all of those records was the wing," says Harry Hyde of Isaac's amazing success. "We had downforce on that salt out there — that was the secret to it. We were breaking records so fast they couldn't get the clock set up! We were going out there to break 60 records, and we broke about 30 of them."

Why did the Nord Krauskopf-owned team end up breaking only half of the records they had originally targeted?

"We could have broken the 60 records but Nord called it off," Harry Hyde insists. "We had talked about cutting the Dodge down and really going for it. He said we should leave some to come back for. And we were going to go back, but we just never got around to it."

Bobby Isaac, poised to strike at Bonneville.

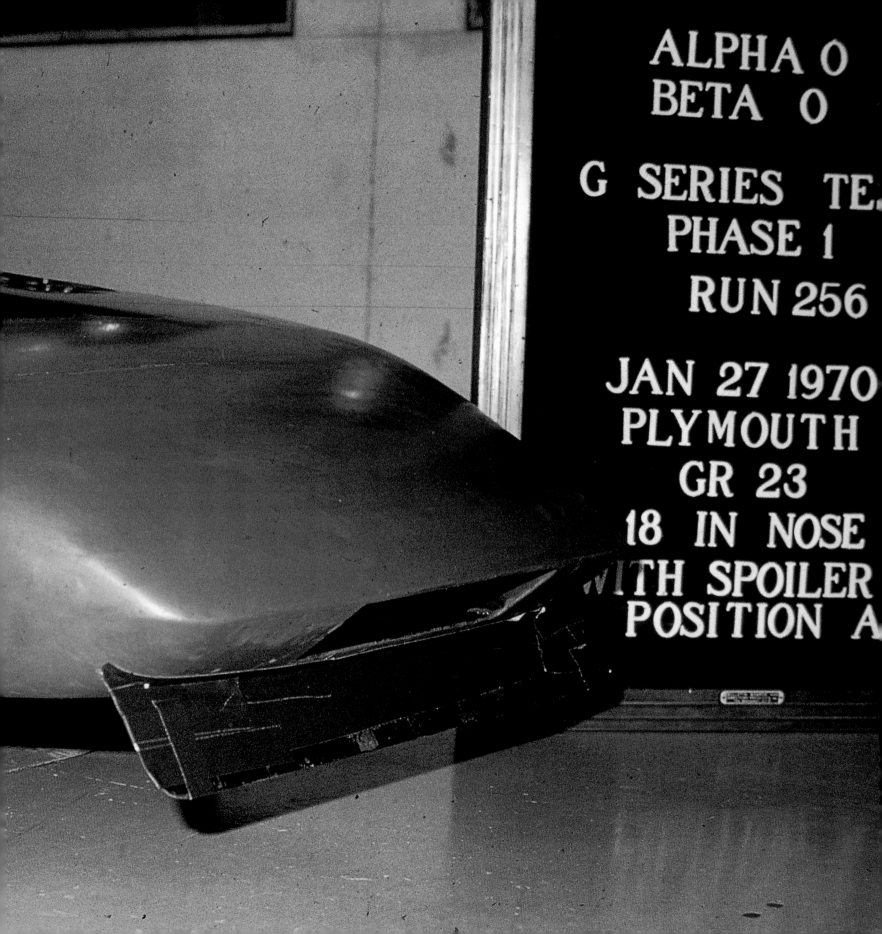

The End of an Era and Wings that Never Were

"Special cars, including the Mercury Cyclone Spoiler, Ford Talladega, Dodge Daytona, Dodge Charger 500 and Plymouth SuperBird shall be limited to a maximum engine size of 305-cubic inches . . ."

With those words from the 1971 NASCAR rule book, it became clear that the last shot in the stock car aerodynamic wars wasn't fired by Chrysler or Ford — it was NASCAR boss Bill France who pulled the trigger. But even before the announcement of new rules to restrict the most exotic stock cars was made, there had been storm clouds on the Grand National racing horizon.

Ford had revised their Torino model for 1970, and with great anticipation took David Pearson, Donnie Allison, and LeeRoy Yarbrough to Daytona for tests in December 1969. They compared the 1969 Torino Talladegas to the new model and came up with distressing results — the 1970 car wasn't as fast as the 1969. Ford was forced to let their top factory drivers compete in the year-old cars throughout the 1970 season, figuring it was better to have a chance at victory in old cars than to lose in new models.

Rumors abounded that Ford had been working on a secret racing project, and those rumors were correct.

In 1969 the head of Ford's racing program, Jacques Passino, knew his drivers would be facing stiff competition from Chrysler's advanced wing cars. Passino assigned designer Larry Shinoda — who had come up with some of Ford's most distinctive designs — to work some magic on the 1970 Torino.

Shinoda faced problems not that far removed from the difficulties his Chrysler colleagues faced with the 1968 Dodge Charger — a backlight that created

 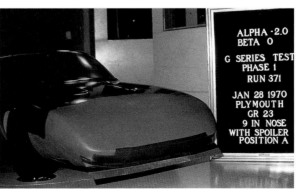

Left and opposite: New beaks for the SuperBird undergoing wind tunnel testing in 1970.

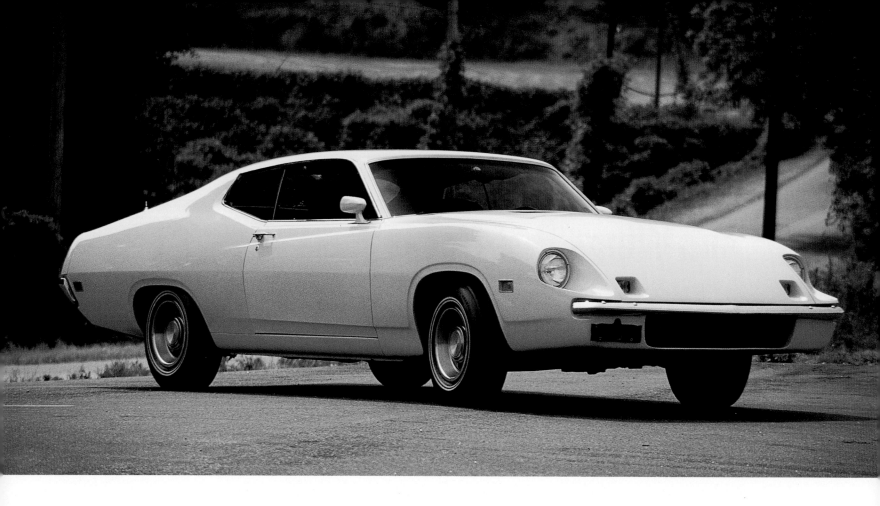

The only surviving Ford King Cobra is owned by NASCAR Winston Cup team owner Bud Moore.

lift and a decidedly non-aerodynamic front grille. Shinoda's solution was the Ford King Cobra.

By late spring 1969, Shinoda's design for the King Cobra was ready for testing at Daytona. A dramatic, sloping front end had been designed to create downforce — and it worked almost too well.

Bud Moore, a famous Ford team owner in the Winston Cup Series, owns the only surviving street prototype of the King Cobra — a yellow rarity with fewer than 700 miles on the odometer. Moore remembers when the first racing King Cobra was taken out for testing at Daytona.

"That front end created so much downforce that it liked to have bent the front axle. That created the problem of rear wheels slipping on the straightaways."

"You talk about a nightmare," Dodge driver Buddy Baker laughs. "That King Cobra was as bad as all the good things that everybody thought about the wing car. They built the car with the sloped nose, but at about 150 MPH the rear wheels would come off the ground!"

Cale Yarborough, who was assigned to test the odd-looking race car that had been built at the Holman-Moody shop in North Carolina, was unable to break the 190 MPH mark — even with a massive Boss 429 motor providing the power. While the front end design created plenty of downforce, the back end still wanted to take off.

"I don't know how it came about, but Ford picked me to test it," says Yarborough, who is more diplomatic in his assessment of the King Cobra's handling problems than his rival Buddy Baker is. "It was a problem to begin with, but we overcame it the more we worked with it. It was a very good car — it handled good, even though it was a big change from what we had been running."

One possible solution to the King Cobra's woes would have been a wing — but the King Cobra program died a sudden death. Semon "Bunkie" Knudson, who, as Ford's president, had been an ardent supporter of the racing program, was dismissed and the racing budget for 1970 was cut dramatically by the new boss, Lee Iacocca. The company had no interest in bearing the financial burden of manufacturing more than 3,000 King Cobras to comply with NASCAR's minimum production rules for Grand National competition. The King Cobra would never have the chance to strike at the SuperBirds and Daytonas.

Had Ford's rivals in the Chrysler camp been concerned about the King Cobra? "No," Special Vehicles Group's Larry Rathgeb emphatically answers. "I had no idea and cared less. I figured kind of like Lee Petty used to say — don't worry about the other guy. You just do the best you can and when you get in competition see how bad off you are. We weren't bad off — we were doing pretty good."

Still, some at Chrysler maintained an interest in what Ford had been up to. "Somebody was down at Daytona when they actually tested the car," reveals Chrysler's John Pointer. "Ford took it to Daytona along with their existing Torino fastback, and somebody was in the press box of the dog track next door with binoculars and a stop watch."

Chrysler hadn't exactly been sitting around waiting to see what Ford would come up with to combat the wing cars. While the automotive magazines wrote about flights of fancy like the Dodge Super Charger — a convertible version of the Daytona built for the auto show circuit that featured a short wraparound spoiler at the car's rear — the real work was going on behind closed doors in the Special Vehicles Group's domain.

With both the Charger and Road Runner scheduled for heavy styling revisions in the 1971 model year, the Chrysler aerodynamics team faced new challenges with the "G-Series" Dodge and Plymouth B-Body cars. The birth process for the new Daytonas and SuperBirds began, once again, with ⅜-scale model tests in the wind tunnel.

"We spent six weeks in the Wichita State University wind tunnel doing development work on the '71 race car," explains aerodynamicist Gary Romberg. "We beat the hell out of the car, developing it to make it just as good as the Daytona."

"On the Daytonas and SuperBirds there's a nice taper back at the C-post and a nice radius at the front," notes test engineer John Pointer of the wing cars' roof line and snout. "The later cars had basically a flat windshield with a sharp corner. The C-

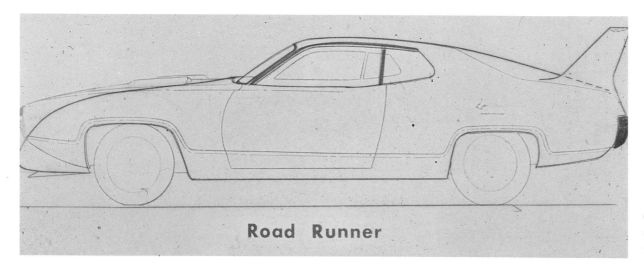

One proposed aero upgrade to the Road Runner body.

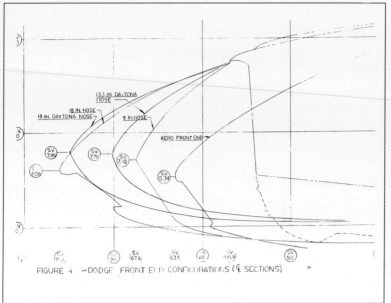
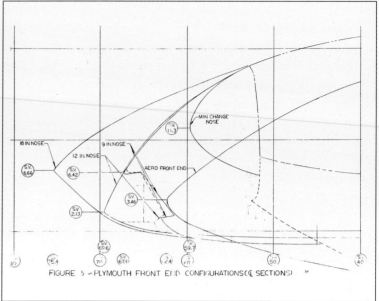

These are the front end designs Chrysler considered for the 1971 wing car models.

post went straight back as well, and the cars were bigger. All of that started to tell against it."

"We had a lot of problems," recalls aerodynamicist Dick Lajoie. "We were in the tunnel and we could never get the drag down on the new one, so we kept putting the old Daytona back in to make sure that the data we were getting was accurate. The first testing was being done in January of 1970, and we spent a lot of effort on this program."

All of that effort was wasted. In November 1970, Chrysler announced drastic cuts in its racing program. Ronnie Householder blamed the cost of racing and the new 1971 NASCAR rules. The first victims of the cutback were the next generation Daytonas and SuperBirds.

What the 1971 wing cars would have looked like has long been the subject of conjecture. Never before revealed outside of the Chrysler Corporation's October 1970 Aerodynamics Research Office study, the results of three months of wind tunnel work (which cost more than $70,000) offer a tantalizing glimpse of what might have been.

The 1971 models, like the first wing cars, were tested with a variety of noses. The Dodge received two types of 18-inch noses, 13.3-inch and nine-inch variations, as well as a fifth, slightly stubbier design, referred to as the "aero front end." On the Plymouth, new front ends of 18, 12, and nine inches were fitted. Engineers also fitted the car with its own "aero front end" design and a rounded "minimum change" nose.

The Chrysler team also studied a variety of roof configurations, including a dramatically swept version of the front windshield. It extended forward more than 26 inches but was pulled back on the sides, resulting in a sleek, semi-circular shape, when seen from above. The design was years ahead of its time.

The Chrysler team also investigated a number of potential backlight plans. They ranged from a shallow drop of the rear glass angle to a beautiful semi-fastback design with completely flush glass.

Wings had made the Daytonas and SuperBirds famous and, fittingly, the most unusual ideas for the 1971 race cars could be found at the back of the cars. Along with a familiar design that featured base stabi-

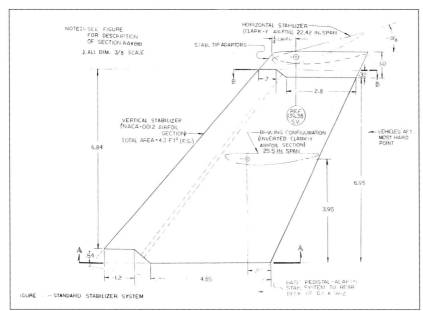 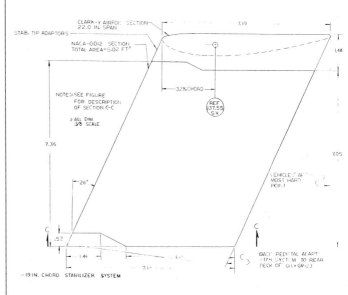

Two varieties of wing configurations that were considered for Chrysler's 1971 model wing cars.

lizers angled at 26 degrees with a 58-inch span Clark-Y airfoil across the top, a radical new design with two wings was conceived. The bi-wing system utilized a 59-inch span across the top with a second, 68-inch wing mounted halfway down to the trunk. Gary Romberg also remembers a tri-wing design not mentioned in the published study that was dubbed "The Red Baron."

The G-Series report concluded that, for the Dodge, the optimum setup was the 18-inch nose without an under-nose spoiler, a semi-fastback backlight, and the stunning bi-wing apparatus mounted in the rear. For the Plymouth, a 12-inch nose was recommended as the most advantageous.

The Special Vehicles Group must have imagined the sight of the 1971 Chrysler cars roaring through the banking at Daytona, rounded snouts in front and two wings keeping the back ends glued to the track. Sadly, this dream never became a reality.

When the 1971 Grand National season began, the NASCAR world looked radically different. In November 1970, shortly after Chrysler announced the cuts to its racing program, Ford virtually quit motorsports entirely. Its racing activities were to be restricted to a limited drag racing and off-road program. Jacques Passino, who was in the middle of a stock car testing session at Riverside International Raceway when the announcement was made, resigned from Ford.

The Chrysler budget slashing that doomed the 1971 wing cars left Ronnie Householder administering a NASCAR program that had withered to one factory-backed Plymouth driven by Richard Petty and one Dodge driven by Buddy Baker. Both cars were fielded out of the Petty Enterprises shops in North Carolina. Petty had hoped to keep Daytona 500 winner Pete Hamilton in the second Petty Enterprises car, but Chrysler preferred the equally talented, hard-charging Baker.

The teams that had lost factory backing but still owned Grand National wing cars had little incentive to race them in 1971 — especially with the NASCAR rules restricting the Daytonas and SuperBirds to tiny 305-cubic-inch engines.

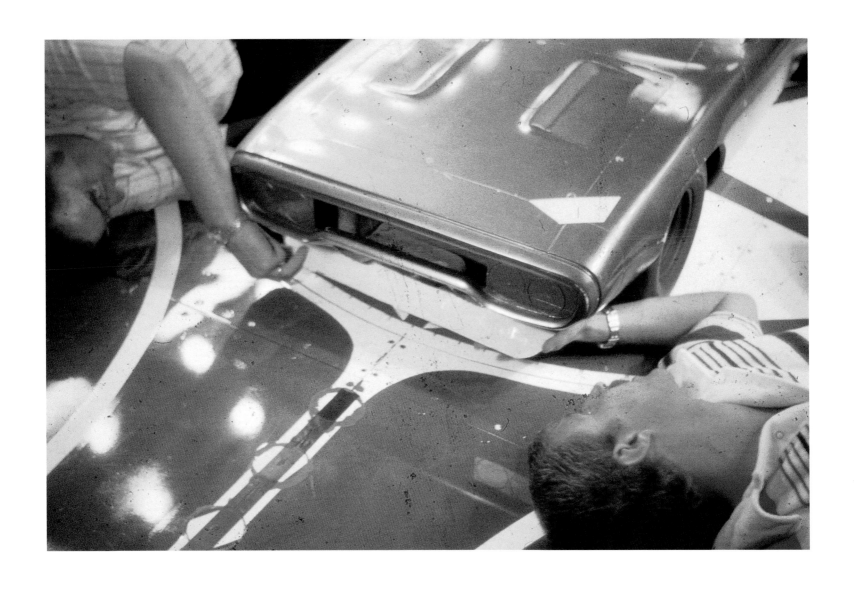

Preceding pages: These rare photographs document the wind tunnel development work performed on scale models representing the 1971 Dodge Charger Daytona (orange models) and Plymouth SuperBird (blue models).

"We never even considered it," says Richard Petty of running a 305-powered SuperBird. "Well, I don't want to say we didn't look at it initially, but we looked at the horsepower of the 305 engine and the horsepower of the 426 and I said, 'Hey, I don't care how good that body is, it can't overcome that.' So we just passed on it."

Virtually banned from NASCAR, the wings did find refuge in other stock car racing series. Drivers competing in the USAC stock car circuit might have doubted the value of wings on USAC's medium-sized tracks, but they soon changed their minds.

Norm Nelson, a top USAC team owner and competitor, was testing Goodyear tires at Milwaukee's famous one-mile track. Chrysler's George Wallace had arranged to do some disc brake testing on Nelson's second car, a SuperBird. The 1971 Plymouth undergoing tire tests was the Nelson team's primary car, which had been set up for maximum speed at the Wisconsin State Fair Park racetrack. The SuperBird had not been set up at all.

"Roger McCluskey was the team's number one driver and Norm Nelson was the number two driver," Wallace recalls. "McCluskey was running a regular

'71 Plymouth in the tire test and they just brought the SuperBird along to put the disc brakes on it for the other test. Both Norm and Roger drove it some and they realized that this car, that didn't have near as good an engine or setup as the one they were tire testing with, was faster."

The wing cars showed their superiority when driven by drivers such as Nelson, McCluskey, and Butch Hartman in the USAC series; Ramo Stott, Jim Vandiver, Iggy Katona, and Bobby Watson in Automobile Racing Club of America events; and Ray Elder, Don White, Dan Gurney, and John Soares in the NASCAR Grand National West series.

The famous drag racing team of Sox and Martin also constructed a SuperBird for quarter-mile racing. The car made infrequent but popular appearances in selected National Hot Rod Association and International Hot Rod Association events as part of the Chrysler "Supercar Clinics" promotion.

With the publication of the 1971 Grand National rules, NASCAR's Bill France figured he had seen the last of the exotic wing cars thundering around his racetracks. He was almost right.

Left and opposite: The wing cars were popular no matter what series they competed in. This USAC SuperBird, raced by the two-car team of owner/driver Norm Nelson and driver Roger McCluskey, was a fan favorite.

A reproduction of the Sox & Martin SuperBird driven by the famous drag racing team at select quarter-mile events.

Many of the teams that lost Chrysler's factory support, including Mario Rossi, were upset. Rossi's red and gold Daytona had been a top contender with Bobby Allison at the wheel throughout 1969 and 1970. Now, with Chrysler having withdrawn its support of his team and NASCAR ruling that his wing car had to use the smallest of engines, Rossi planned revenge.

When the Grand National cars arrived in Florida for the 1971 Daytona 500 there were no wing cars — except for one red and gold Daytona with the legend "305 C.I." painted across the hood. Mario Rossi was going to try to win NASCAR's biggest race with the car Bill France thought he had legislated out of existence.

Bobby Allison had moved on after the 1970 season, so Rossi hired Dick Brooks — who had made several strong runs in a SuperBird in 1970 — to pilot his Dodge in the 500-mile race. Chrysler couldn't resist making arrangements to fund Rossi's efforts to put another wing car in Daytona's victory lane.

"Rossi's the only one who really paid a whole lot of attention to what I was doing," Brooks recalls. "When the factories left and Bobby went off and got another job, this thing came up about running the wing car again. I got to talking to Rossi about it and I said, 'Well, it sure sounds good to me.' I didn't have a job and neither did Rossi, and Chrysler was paying for it. So we got together and went out to California to meet with Keith Black for the motor stuff."

In an era where brute force ruled the roost, to gain enough power to run competitively with just 305-cubic inches was going to require a very special motor. Keith Black, a legendary engine builder, could provide that power if anybody could.

"From the beginning it didn't seem like it was going to work," Brooks continues. "But Keith Black's organization had pretty good engine builders. They'd used smaller engines and they knew the motor. It was a whole lot lighter — a small block, light rods, light pistons. It was a little bitty motor they nicknamed the 'lunchbox.' When somebody walked up they'd say, 'Hey, man, somebody left their lunchbox under your hood!'"

During his first practice sessions for the Daytona 500, Brooks discovered a new characteristic of the "lunchbox." Having raced with a Hemi under his hood, Brooks was used to an engine operating in the 7,000 RPM range. The Keith Black powerplant went just a little higher than that as it tried to generate force to match the big motors.

"I got a really good line through turns one and two," Brooks recalls. "I looked down and it was turning about 9,800 RPM about halfway down the straightaway! I just couldn't keep my eyes off the tach! I was just watching it and at the same time wanting to get my foot out of it. You're just thinking of the thing coming apart and cutting you all to pieces, but I got it a little over 10,000 RPM.

"It sounded funny," Brooks remembers. "In those days with those big old thumpers this had a little more of a whining sound. It sounded funny, but that little sucker sure would run. Running by itself it didn't do anything, but what a hoss it was when the race started!"

That it was. To the joy of the Chrysler world and to the shock of Bill France and NASCAR, by lap 60 Dick Brooks managed to get the number 22 Daytona into first place. "They tell me the drunks fell out into the aisles!" Brooks laughs, remembering the race. "It was great while I was leading. We were only credited with leading 18 laps but we led the pack for a long, long time."

Unfortunately, Brooks's miracle in the making was not to be. Pete Hamilton, who had won the 1970 Daytona 500 in a SuperBird, collided with Brooks, ending the Charger Daytona's chance to win one final Grand National race.

The last hurrah of the NASCAR wings. Dick Brooks driving Mario Rossi's Charger Daytona in the 1971 Daytona 500 with the "lunchbox" purring under the hood.

The wings were gone from Grand National racing, but they flew elsewhere. Gary Bettenhausen prepares to race in a USAC event at Pocono, Pennsylvania (above), while John Soares's SuperBird kicks up a rooster tail of dust in California (opposite). This rare glimpse of a wing car dirt track racing was captured at the Sacramento Fairgrounds in May 1970 at a Grand National West event.

"He and I got together," Hamilton says. "We took each other out. I ended up falling out and he lost a lap. That was the swan song for that good old car."

Dick Brooks finished seventh in the 1971 Daytona 500. It was the last time a Chrysler wing car raced in the NASCAR Grand National Series.

"We really wanted that car to win," recalls Chrysler's Gary Romberg. "It would have been neat for Chrysler, but to really stick it to the NASCAR people — that would have been fun to do. If we could have won with that car they probably would have put a 105-cubic-inch limit on it! The message from NASCAR was 'Take those cars off the track.'"

Why was NASCAR so determined to remove the exotic wing cars from Grand National racing?

Larry Rathgeb believes France made the prohibitive rules because he was losing control. "He couldn't cope with the factory representatives there at his races. He had to get rid of them. Not only that, but we had control over the drivers because they were under contract to the manufacturers. So he finally pitched us all out."

"It would have been nice to continue," says Dodge public relations director Frank Wylie. "We spent a lot of time trying to get a fair shot, and then when we caught up they changed the rules. We felt strongly about that, and that's why we essentially went elsewhere. We put a greater emphasis on drag racing."

"Detroit was controlling NASCAR," believes Richard Petty. "Every week they came down with a new gimmick on a new car or something, and NASCAR couldn't keep up with it. They decided they didn't want to keep up with it. They said, 'Hey — it's our ball game. We want you to play with our bat and our ball, and we're going to tell you what that bat and ball are going to look like instead of you telling us what they're going to look like.'

"It started in 1949 as a stock appearing series," Petty continues, "and here were these cars with wings and these sloped noses and these sloped back ends. If they didn't write Plymouth or Dodge or Ford on the side people wouldn't even know what they were! So they got to be prototype cars, and Bill France knew it was getting away from the stock deal and he didn't want that."

"I think banning exotic cars was the right thing to do," affirms Darlington International Raceway president Jim Hunter. Hunter covered the Grand National series as a reporter during the aero wars. "NASCAR's philosophy of equating the rules from race to race — the teams don't like it and the manufacturers don't like it, but in the long haul it's been one of the things that has made NASCAR racing so popular. The competition among the different makes has been kept as close as possible, so that no one make winds up with a decided advantage."

Although stock car racing continued to grow and flourish without the wing cars, never again would the pace of factory innovation and development match the feverish pitch of the competition found on the NASCAR Grand National superspeedways.

One can't help but wonder how fast the wing cars could have gone if they had continued to evolve.

"If you could lower the wing car down, put today's tires on it, and evaluate the difference in horsepower over the years of development, you'd have a race car that would run 235 MPH at Talladega," speculates Buddy Baker.

"You know, NASCAR was right in parking that car," states Harry Hyde, crew chief of the K&K Insurance Daytona with which Bobby Isaac won the 1970 NASCAR championship. "At the time I thought it was a cruel act, but we didn't know how to be careful. We would have had them going so fast that it wouldn't even have been safe to sit in the grandstand."

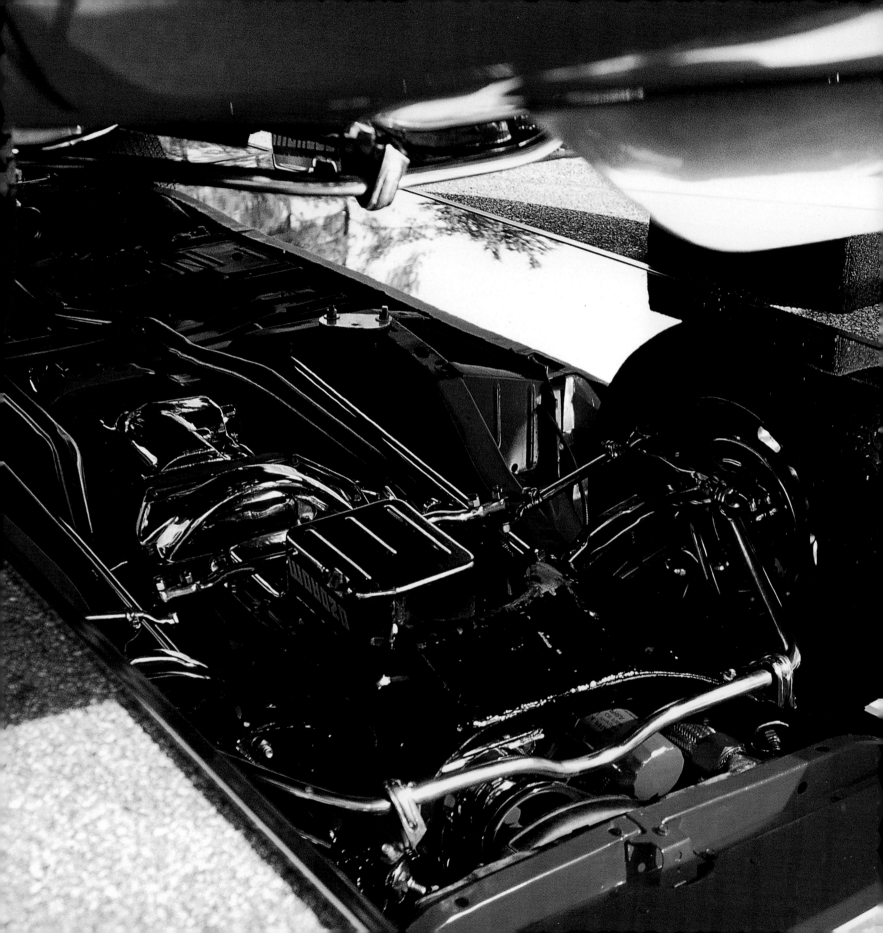

case like a lot of these cars are when they come in here, I'd just as soon do this complete restoration to it and make the car last longer."

The biggest challenge in the restoration procedure is finding out what is and isn't correct for a wing car. Gathering information about original wing car construction is an ongoing process.

"The only way to know something is perfectly correct for a particular car is to find it that way on that car," Ferro explains. "We'll take an individual part, and we'll slowly start taking layers off, reversing time to find out how that part originally was. If you have a 4,000-mile Superbird on your hands, you try to get as much information as you can off that. The problem is you might find another one a little different. The reason is that Chrysler used several vendors, and whoever bid the cheapest for that part got the job. You have to be knowledgeable enough about the car to know what you're looking at and say, 'Wait a minute — somebody changed this along the way,' but also be open minded enough to say, 'This isn't known to be correct but it sure looks like it came that way.' The ultimate place to find good sources is original cars."

While excavating the innards of wing cars Ferro has come across all kinds of oddities, like tic-tac-toe games or "Help me!" inscribed in the inside of fuel tanks, or messages encoded in other hidden areas from one assembly line worker to a buddy further down the line.

To resolve questions of authenticity, restorers frequently keep in contact with each other.

"If I have a question, no matter how stupid it might seem to me or you, it's important to the car," Ferro points out. "If I know a guy who has an original he's working on, then we'll talk all the time. Some people try to keep secrets, and they're usually not around that long. We're all really just trying to help the hobby — there's no rocket science here. There's no way to learn a lot unless information is changing hands."

Finding parts for cars that are decades old can be a real headache. "If you need a part you make it,"

says Ferro. The other option is hunting down parts from other restorers. "It's hours and hours and hours of talking to people on the phone and searching," he explains. "There's very few wing car-specific parts that you can get reproductions of, but I haven't had anything come through here that we couldn't fix."

Ferro's Totally Auto finishes about ten complete restorations each year, and the waiting list to bring a car to Ferro's attention is over a year long. The cost for an immaculate restoration averages in the $50,000 range. If that sounds like a lot to pay for a Daytona restoration, consider that the asking price for a Daytona in superior shape can vary from $70,000 to $145,000 in the volatile collector's market, with SuperBirds ranging from $40,000 to $70,000 — and both cars become rarer with each passing year.

But of all the work on muscle cars and special autos Dave Ferro does, his wing car work has special meaning.

"I think they're great. It's got to be the most recognizable car ever made," Dave explains. "I've always loved those cars, and I think they're one of the few Mopars that when they command the high prices, they're worth it. They are a very special vehicle."

While many wing car owners turn to a professional like Dave Ferro for help with their restorations, some owners prefer to do the work themselves — and the conditions of the cars they start with can rival the worst examples hauled into Ferro's shop.

"The driver's side door was wrecked, and a pole had fallen down across the top and it was pushed right down to the tops of the seats," Don Snyder recalls of his SuperBird's condition when he took possession of it. "Plus the hood was destroyed. The nose was gone, but the engine was all still there and the original driveline."

Snyder had first seen the car that would eventually become his beautiful Tor-Red SuperBird at an auction in Syracuse, New York, back in 1972. "I decided not to buy the car then because it was kind of rough, and the car disappeared," says Snyder. "A few years later I found it in a junkyard all wrecked. I

Don Snyder's Plymouth SuperBird, a beautiful car Snyder lovingly restored himself.

It took some detective work, but Tim Kirkpatrick finally discovered his Daytona's former identity. It had once been a pace car at a New York racetrack (left). Today, it's a stunning example of Dodge's finest (below).

pictures of the car pacing the field at Perry Speedway. "Eventually," explains Kirkpatrick, "they had hidden the checkers on the roof by spraying a vinyl roof on it. It was pretty neat to be able to find out where it had come from."

Today Kirkpatrick's immaculate red Daytona shows no signs of its racetrack heritage — although that could very well change in the future.

"I'm having some mixed thoughts right now of maybe doing it as a pace car, with its original pace car look," says Kirkpatrick. "That's how it was, and that's what it was intended for when it was originally done."

While Kirkpatrick's Daytona would certainly look wilder in its original pace car colors, changes to John Pappas's SuperBird could only make it look tamer. In a tribute to the legendary Chrysler drag racing team of Sox & Martin, Pappas turned an ordinary white Plymouth into a replica of one of the greatest wing legends — the Sox & Martin SuperBird that made several popular appearances at National Hot Rod Association events and match races in 1970.

"I built a lot of model cars, and Sox & Martin were always my heroes," Pappas explains. His wing car's fate was sealed when he finished a model of the famous red, white, and blue drag racing SuperBird.

"I finished the model, and I was looking at my SuperBird," Pappas smiles. "I thought, 'My car's white — it wouldn't be too hard to add the red and the blue.'"

Pappas had found his SuperBird some years before, and owning one of the wing cars was the realization of a lifelong dream.

"I was seven or eight and I used to ride to church with my mom and dad and grandmom," Pappas remembers. "We'd go past all of the car dealers in Detroit, and one day a Chrysler-Plymouth dealer had a yellow 'Bird and an orange 'Bird, right when the cars first came out. I said, 'Mom! Dad! Aren't those cars neat?' And they said, 'Those are the ugliest cars I've ever seen!' To me, I knew right from that point that someday, no matter what, I was going to find one. I knew I had to have one someday.

"In 1981 I really started looking," Pappas continues, "and we found this car down in Columbia, South Carolina. It was on consignment to a little used car lot by the original owner. The car had 26,000 miles on it. We went down there and they were asking $7,500 for it. We offered them $5,800 and the next thing we knew we were driving back to Michigan in a 440 SuperBird!"

Several years later, Pappas made the decision to make his SuperBird into a rolling Sox & Martin shrine.

"Years went by, and I started accumulating pieces to make the change," John says. "I didn't know what to do about the hood — that was a problem. I didn't want to leave the hood flat because it didn't look all that aggressive, but I knew I wasn't going to cut up my original hood to make it look like the hood with a Six-Pack scoop on their car. I wasn't going to cut a hole in my hood — an irreplaceable SuperBird hood. You can't do that!

"Finally, a guy in Texas starting reproducing a lot of body parts for the SuperBird, including the extensions to make a 1970 Coronet hood into a SuperBird hood," Pappas recounts. "When I saw that, I said, 'Wow! This is the way I can do it!' So I got one of his pieces, and I went over to my painter and handed him all of the parts and the car, went off to the graphics place and had them do all of the graphics, and I bought the last set of Keystone wheels in

Left and opposite: Sox & Martin fan John Pappas's recreation of one of the most legendary wing cars is hugely popular everywhere it goes.

the city of Detroit. About a month after that, the car was done."

John Pappas's Sox & Martin replica is one of the most popular wing cars in existence today. Everything about it — from the massive hood scoop to the competition graphics, from the high stance of the rear end resting on wide tires to the Keystone wheels — leaves no doubt that this is a machine built for speed. Pappas can park his SuperBird and, almost instantly, a curious crowd will materialize.

"In the last three or four years I've probably put four or five hundred miles on it on the street," Pappas says. "Other than that, it just goes a quarter mile at a time at nostalgia races. We've been to a lot of shows, and it's neat because a lot of people have never seen the Sox & Martin SuperBird. People bring their kids up and have their pictures taken next to the car. It's just neat to bring a smile to people's faces and make them happy with a car that brings back memories of NHRA racing like it was back in 1969 and 1970."

Like much of wing car history, what happened to the real Sox & Martin SuperBird remains a mystery.

"I still don't know if this is true," Pappas begins, "but their cars were supposedly given back to Chrysler. Because they were race cars, they could not be sold to the public and therefore had to be de-

It is a rare sight, but unrestored SuperBirds and Daytonas do still exist. This Daytona (bottom) is especially mysterious because it has R/T strobe paint and trim.

stroyed. But I've heard rumors that there is one car in Louisiana and I've heard another rumor there's a car in the northeastern part of the country — but I have no proof they exist.

"How do these rumors start?" Pappas wonders. "Did somebody just have a SuperBird that was kind of like mine, that was painted up back then, and it became legend that it was the real car? It's hard to say where people get these ideas."

Indeed, rumors of forgotten wing cars play a big part in conversations among wing car collectors, with reported sightings of unrestored wing cars languishing in a barn or back lot speeding through the enthusiast grapevine. Fans of these cars drive country roads with one eye cocked towards every field or backyard in the hopes of seeing a wing reaching skyward or a rounded snout peeking around a corner.

Even more excitement can be generated by the discovery of a car that is undocumented. Just such a car sits in the wilds of central Pennsylvania, open to the elements and decaying with each passing season. It appears that it is a Dodge Charger Daytona based on the 1970 Charger R/T — but no such wing car was ever sold. The body, which features the R/T side trim and "strobe" stripe paint, has been examined on the sly by more than one wing car expert — and the claim is that this is indeed a factory-built car, not an owner-modified 1970 R/T. Rumor has it that the car may have been built by Creative Industries in an attempt to entice Chrysler into building a 1970 Daytona model. Overtures from wing car conservationists to purchase this unique vehicle have fallen on deaf ears. The car's owner is apparently content to let this Dodge oddity rot away.

The Daytona and SuperBird are the focus of two organizations dedicated to the preservation of wing cars: The Daytona-SuperBird Auto Club and the Winged Warriors. Aside from social functions, the clubs can provide valuable services to their members, as Tim Kirkpatrick can attest.

Like many Daytona owners, Kirkpatrick knew that the Dodge factory "build sheet" that actually belonged to his car was not the same sheet he found attached to his Daytona's rear seat unit. It seems that when the Daytonas were being converted from regular Chargers back in 1969, the back seats — and the sheets listing detailed information about the car's origins and equipment — were simply heaped in a pile. As a result, various back seats and build sheets were randomly reinserted in the wing cars.

One day Kirkpatrick received a call from a Daytona owner in California who had tracked

Twenty-five years after their first race, the wing cars returned to Talladega to commemorate a unique period in automotive history.

**Right and opposite:
The Aero Warrior
Reunion in Talladega,
Alabama, May 1994.**

Kirkpatrick down through one of the wing car clubs. This man's Daytona had the build sheet that belonged in Kirkpatrick's Daytona, and soon car and build sheet were reunited.

Large gatherings of wing cars become rarer with each passing year. One of the largest ever was the Aero Warrior Reunion held in Alabama in May 1994. Activities centered around the Talladega Superspeedway, where the Daytona made its debut with Richard Brickhouse's victory in 1969 and where Pete Hamilton won twice in a Petty Enterprises SuperBird in 1970.

Appropriately, both Brickhouse and Hamilton attended the event, as did many other wing car luminaries, including drivers Bobby Allison and Buddy Baker, Chrysler aerodynamicist Gary Romberg, Chrysler engineering team mechanic Larry Knowlton, and the man who oversaw the wing cars' NASCAR performance, Chrysler's Larry Rathgeb. Joining in the excitement were dozens of wing car owners who, of course, brought their cars, as well as a contingent of Ford Torino Talladega and Mercury Cyclone Spoiler II owners. Two long lines of aero warriors made a parade lap around Talladega Superspeedway before the NASCAR Winston Cup Series race, reminding thousands of fans of a glorious bygone era.

Coordinating the Aero Warrior Reunion and putting all of the pieces together was a huge undertaking, but it was a labor of love for Alabama wing car enthusiast Tim Wellborn.

"My father had always had Dodges, and being near Talladega we went to that first race in 1969," Wellborn explains. "To see those cars for the first time, and to hear the sounds when you were fourteen years old — well, that would put it in your heart forever. I was lucky to get to see that."

Wellborn never did get the wing cars out of his system. After directing a successful 1988 gathering of wing cars at Talladega, Wellborn began to plan an elaborate 25th anniversary celebration for 1994 that would not only include the cars, but also the drivers and Chrysler personnel linked to wing car history.

As the Aero Warrior Reunion approached, Wellborn worked closely with Don Naman, the director

of the International Motorsports Hall of Fame and Museum at Talladega. The president of Talladega Superspeedway, Grant Lynch, also offered his resources. Since a Daytona won the first Grand National race at Talladega, the idea of a celebration honoring both the track and the wing cars 25 years later was an appealing one.

"We spent many hours together working this thing up," Wellborn says, "It was really gratifying to see it all come together and know that this may have been the elite meet in the history of the cars. I think the quality of this meet was higher because of the drivers that we had speaking to the group and the fact that the track was celebrating it, too.

"We're definitely going to do a 30th reunion," Wellborn promises. "My love for the wing cars is something I'll always have. They were special in their time, and they'll always be special to anyone who really looks at them. They were a true race car."

Tim Wellborn's dedication to the Daytona and Superbird is characteristic of the fascination and loyalty these cars are capable of generating.

"The memories that we have of these cars will last a lifetime," John Pappas explains. "It's indescribable, the looks that these cars bring. You answer a lot of questions, but it's been worth every minute of it. I'll probably die owning this car — it's been with me too long. I more or less grew up with it."

Those who raced in the Daytonas and SuperBirds have equally fond feelings about the wing cars.

"I hated to see the wing cars go," says Buddy Baker. "They had so much potential, and they were getting better and better as we ran them from the knowledge we were finding out about them. It was just awesome."

"I started in racing right after World War II and I've been in racing ever since," says Bobby Isaac's crew chief, Harry Hyde. "But I can't forget those old days. It was a good era to go through, and a good time to be living."

"There's never been anything like them and there never will again," says Dick Brooks, summing up the appeal of these wild cars.

Tim Wellborn's immaculate SuperBird resides in Talladega's museum for all to enjoy. With the exception of oil and spark plug changes, this wing car is exactly the same as it was when it left Chrysler's Lynch Road assembly plant.

Dodge Charger Daytonas, Plymouth SuperBirds, and their Ford competition return to the high banks of Talladega Superspeedway.

The Chrysler wing cars are vehicles that pushed the limits, and their story is one unlike any other in automotive history. But perhaps the best assessment of the Dodge Charger Daytona and Plymouth SuperBird comes from the man responsible for their NASCAR speedway success, Larry Rathgeb.

"In the history of racing, I'd say we achieved the ultimate," Rathgeb reflects. "I think we did a fine job, and we felt a tremendous amount of accomplishment. All we did was try to join the aerodynamics with the handling process and make it all work. It's not me or any one person — it was a whole group of people. It was a whole organization and I think that everybody at Chrysler all felt they were a part of it."

Larry Rathgeb pauses, and then smiles.

"We never thought it was going to be anything so big, that people would remember them so well today."

Following pages: The ultimate dream of Dodge Charger Daytona and Plymouth SuperBird collectors came true for Bill Stech — he owns a superb example of each.

Appendix

VEHICLE PRODUCTION INFORMATION FOR THE DODGE CHARGER DAYTONA

As is the case with many models from the muscle car era, the exact number of Charger Daytonas built is uncertain. Published figures total between 501 and 507 Daytonas built, ordered, and shipped to dealers in the United States, with the numbers 503 and 505 being the most frequently cited. The vast majority were equipped with the 440 4bbl. engine, although the precise number equipped with the 426 Hemi powerplant is another figure open to debate. Hemi count figures range from 32 and 37 up to a high of 70 according to various sources. It has been reported that approximately 50 additional Daytonas made their way to Canada, in which case the total production figure for the Charger Daytona would be in the vicinity of 550.

The Dodge Charger Daytona was available in the following colors:

Silver Metallic, Light Blue Metallic, Bright Blue Metallic, Medium Blue Metallic, Light Green Metallic, Medium Green Metallic, Bright Green Metallic, Dark Green Metallic, Bright Turquoise Metallic, Bright Red, Red, Light Bronze Metallic, Copper Metallic, Dark Bronze Metallic, Hemi Orange, Eggshell White, Black, Yellow, Cream, Light Gold Metallic, and Special Order Paint.

It is believed just over 350 Dodge Charger Daytonas exist today.

VEHICLE PRODUCTION INFORMATION FOR THE PLYMOUTH SUPERBIRD

While no one can seem to agree on the number of Daytonas built, the figures vary even more wildly when it comes to its winged cousin, the Plymouth SuperBird. Chrysler memos of September 1969 show that the Sales Programming staff was preparing to handle 1,920 winged Plymouths for 1970, but published figures say as many as 2,783 were built. The current figure generally accepted is 1,935 SuperBirds built and shipped to United States dealers, with anywhere from 34 to 47 allegedly heading towards Canada. The engine option question is again a sticky one, although the most frequently seen numbers report 135 426 Hemi SuperBirds and 716 440 Six-Pack editions, with the remainder powered by 440 4bbl. motors.

It is believed that just over 1,000 Plymouth SuperBirds exist today.

THE DODGE CHARGER DAYTONA AND PLYMOUTH SUPERBIRD IN COMPETITION

While the performance records of the wing cars in their 30 NASCAR Grand National competition appearances are relatively comprehensive, the United States Auto Club (USAC) and Automobile Racing Club of America (ARCA) wing car results are open to interpretation. Often race results recorded just the manufacturer, not specific models. Thanks to Donald Davidson at USAC, Ron Drager at ARCA, and Doug Schellinger of the Daytona-SuperBird Auto Club, the results for USAC and ARCA races which follow the Grand National statistics are as accurate as possible.

NASCAR GRAND NATIONAL WING CAR COMPETITION RESULTS

Talladega 500
Alabama International Motor Speedway (2.66 mi.)
Talladega, AL, September 14, 1969

The Top Five:
1. Richard Brickhouse, #99 Nichels Engineering Daytona
2. Jim Vandiver, #3 Ray Fox Dodge Charger
3. Ramo Stott, #14 Bill Ellis Dodge Charger
4. Bobby Isaac, #71 Nord Krauskopf Daytona
5. Dick Brooks, #32 Dick Brooks Plymouth

Other Wing Car Finishes:
Brickhouse and Isaac were the only two wing car entries.

Pole Winner:
Charlie Glotzbach, #88 Nichels Engineering Daytona, 199.466 MPH
(Glotzbach left with the PDA boycott and Bobby Isaac started the Alabama 500 from the pole postion since he was the fastest of the first round qualifiers who stayed to compete. Isaac's qualifying speed was 196.386 MPH.)

National 500
Charlotte Motor Speedway (1.5 mi.)
Charlotte, NC, October 12, 1969

The Top Five:
1. Donnie Allison, #27 Banjo Matthews Ford
2. Bobby Allison, #22 Mario Rossi Daytona
3. Buddy Baker, #6 Cotton Owens Daytona
4. Charlie Glotzbach, #99 Nichels Engineering Daytona
5. David Pearson, #17 Holman-Moody Ford

Other Wing Car Finishes:
11. James Hylton, #48 Hylton Engineering Daytona
30. Dave Marcis, #30 Milt Lunda Daytona
37. Richard Brickhouse, #88 Bill Ellis Daytona
41. Bobby Isaac, #71 Nord Krauskopf Daytona

Pole Winner:
Cale Yarborough, #21 Wood Brothers Mercury, 162.162 MPH

American 500
North Carolina Motor Speedway (1.017 mi.)
Rockingham, NC, October 26, 1969

The Top Five:
1. LeeRoy Yarbrough, #98 Junior Johnson Ford
2. David Pearson, #17 Holman-Moody Ford
3. Buddy Baker, #6 Cotton Owens Dodge Charger
4. Dave Marcis, #30 Milt Lunda Daytona
5. John Sears, #4 L.G. DeWitt Ford

Other Wing Car Finishes:
23. Charlie Glotzbach, #99 Nichels Engineering Daytona
31. Richard Brickhouse, #88 Bill Ellis Daytona
39. Bobby Allison, #22 Mario Rossi Daytona

Pole Winner:
Charlie Glotzbach, #99 Nichels Engineering Daytona, 136.972 MPH

Texas 500
Texas International Speedway (2 mi.)
College Station, TX, December 7, 1969

The Top Five:
1. Bobby Isaac, #71 Nord Krauskopf Daytona

2. Donnie Allison, #27 Banjo Matthews Ford
3. Benny Parsons, #18 Russ Dawson Ford
4. James Hylton, #48 Hylton Engineering Daytona
5. Dick Brooks, #32 Dick Brooks Plymouth

Other Wing Car Finishes:
6. Ray Elder, #96 Elder Daytona
7. Jack McCoy, #7 Ernie Conn Daytona
8. Buddy Baker, #6 Cotton Owens Daytona
9. Dave Marcis, #30 Milt Lunda Daytona
23. Bobby Allison, #22 Mario Rossi Daytona
33. Richard Brickhouse, #99 Nichels Engineering Daytona

Pole Winner:
Buddy Baker, #6 Cotton Owens Daytona, 176.284 MPH

Motor Trend 500
Riverside International Raceway (2.62 mi. road course)
Riverside, CA, January 18, 1970

The Top Five:
1. A.J. Foyt, #11 Jack Bowsher Ford
2. Roger McCluskey, #1 Norm Nelson SuperBird
3. LeeRoy Yarbrough, #98 Junior Johnson Ford
4. Donnie Allison, #27 Banjo Matthews Ford
5. Richard Petty, #43 Petty Enterprises SuperBird

Other Wing Car Finishes:
6. Dan Gurney, #42 Petty Enterprises SuperBird
7. Neil Castles, #06 Castles Daytona
13. Bobby Allison, #22 Mario Rossi Daytona
14. Dave Marcis, #30 Marcis Daytona
24. Ray Elder, #96 Fred Elder Daytona
29. Bobby Isaac, #71 Nord Krauskopf Daytona
35. James Hylton, #48 Hylton Engineering Daytona
39. Don White, #93 Nichels Engineering Daytona

Pole Winner:
Dan Gurney, #42 Petty Enterprises SuperBird, 112.060 MPH
(A 113.310 MPH run by Parnelli Jones in the Wood Brothers Mercury was disallowed due to the use of ineligible tires.)

125-Mile Qualifying Race One for 1970 Daytona 500
Daytona International Speedway (2.5 mi.)
Daytona Beach, FL, February 19, 1970

The Top Five:
1. Cale Yarborough, #21 Wood Brothers Mercury
2. Bobby Isaac, #71 Nord Krauskopf Daytona
3. LeeRoy Yarbrough, #98 Junior Johnson Ford
4. Donnie Allison, #27 Banjo Matthews Ford
5. Pete Hamilton, #40 Petty Enterprises SuperBird

Other Wing Car Finishes:
6. Richard Petty, #43 Petty Enterprises SuperBird
7. Dick Brooks, #32 Dick Brooks SuperBird
8. Ramo Stott, #7 Stott SuperBird
12. Dr. Don Tarr, #37 Tarr Daytona
18. Neil Castles, #06 Castles Daytona

Pole Winner:
Cale Yarborough, #21 Wood Brothers Mercury, 194.015 MPH

125-Mile Qualifying Race Two for 1970 Daytona 500
Daytona International Speedway (2.5 mi.)
Daytona Beach, FL, February 19, 1970

The Top Five:
1. Charlie Glotzbach, #99 Nichels-Goldsmith Daytona
2. Buddy Baker, #6 Cotton Owens Daytona
3. Bobby Allison, #22 Mario Rossi Daytona
4. Tiny Lund, #55 McConnell Daytona
5. Richard Brickhouse, #14 Bill Ellis SuperBird

Oher Wing Car Finishes:
6. Ray Elder, #96 Fred Elder Daytona
25. Buddy Arrington, #5 Arrington Daytona
28. Talmadge Prince, #78 Hodges-Prince Daytona

Note:
Prince was killed in a crash which occurred on lap 18.

Pole Winner:
Buddy Baker, #6 Cotton Owens Daytona, 192.624 MPH

Daytona 500
Daytona International Speedway (2.5 mi.)
Daytona Beach, FL, February 22, 1970

The Top Five:
1. Pete Hamilton, #40 Petty Enterprises SuperBird
2. David Pearson, #17 Holman-Moody Ford
3. Bobby Allison, #22 Mario Rossi Daytona
4. Charlie Glotzbach, #99 Nichels-Goldsmith Daytona
5. Bobby Isaac, #71 Nord Krauskopf Daytona

Other Wing Car Finishes:
6. Richard Brickhouse, #14 Bill Ellis SuperBird
8. Ramo Stott, #7 Stott SuperBird
10. Dave Marcis, #30 Marcis Daytona
11. Ray Elder, #96 Fred Elder Daytona
12. Neil Castles, #06 Castles Daytona
13. Tiny Lund, #55 McConnell Daytona
19. Dick Brooks, #32 Dick Brooks SuperBird
27. Buddy Baker, #6 Cotton Owens Daytona
29. Buddy Arrington, #5 Arrington Daytona
36. Dr. Don Tarr, #37 Tarr Daytona
39. Richard Petty, #43 Petty Enterprises SuperBird

Pole Winner:
Cale Yarborough, #21 Wood Brothers Mercury, 194.015 MPH

Carolina 500
North Carolina Motor Speedway (1.017 mi.)
Rockingham, NC, March 8, 1970

The Top Five:
1. Richard Petty, #43 Petty Enterprises SuperBird
2. Cale Yarborough, #21 Wood Brothers Mercury
3. Dick Brooks, #32 Dick Brooks SuperBird
4. Bobby Allison, #22 Mario Rossi Daytona
5. Pete Hamilton, #40 Petty Enterprises SuperBird

Other Wing Car Finishes:
6. Dave Marcis, #30 Marcis Daytona
14. Bobby Isaac, #71 Nord Krauskopf Daytona
17. Charlie Glotzbach, #99 Nichels-Goldsmith Daytona
22. Dr. Don Tarr, #37 Tarr Daytona
32. Neil Castles, #06 Castles Daytona
33. Buddy Baker, #6 Cotton Owens Daytona

Pole Winner:
Bobby Allison, #22 Mario Rossi Daytona, 139.048 MPH

Atlanta 500
Atlanta International Raceway (1.522 mi.)
Hampton, GA, March 29, 1970

The Top Five:
1. Bobby Allison, #22 Mario Rossi Daytona
2. Cale Yarborough, #21 Wood Brothers Mercury
3. Pete Hamilton, #40 Petty Enterprises SuperBird
4. LeeRoy Yarbrough, #98 Junior Johnson Ford
5. Richard Petty, #43 Petty Enterprises SuperBird

Other Wing Car Finishes:
10. Dr. Don Tarr, #37 Tarr Daytona
20. Buddy Baker, #6 Cotton Owens Daytona
26. Dick Brooks, #32 Dick Brooks SuperBird
28. Bobby Isaac, #71 Nord Krauskopf Daytona
31. Dave Marcis, #30 Marcis Daytona
38. Charlie Glotzbach, #99 Nichels-Goldsmith Daytona
39. Neil Castles, #06 Castles Daytona

Pole Winner:
Cale Yarborough, #21 Wood Brothers Mercury, 159.929 MPH

Alabama 500
Alabama International Motor Speedway (2.66 mi.)
Talladega, AL, April 12, 1970

The Top Five:
1. Pete Hamilton, #40 Petty Enterprises SuperBird
2. Bobby Isaac, #71 Nord Krauskopf Daytona
3. David Pearson, #17 Holman-Moody Ford
4. Benny Parsons, #72 L.G. DeWitt Ford
5. Cale Yarborough, #21 Wood Brothers Mercury

Other Wing Car Finishes:
6. Freddy Fryar, #14 Bill Ellis SuperBird

7. Richard Petty, #43 Petty Enterprises SuperBird
9. Neil Castles, #06 Castles Daytona
12. Buddy Baker, #6 Cotton Owens Daytona
13. Dick Brooks, #32 Dick Brooks SuperBird
29. Bobby Allison, #22 Mario Rossi Daytona
31. Charlie Glotzbach, #99 Nichels-Goldsmith Daytona
33. Jim Vandiver, #31 Vandiver Daytona
38. Dr. Don Tarr, #37 Tarr Daytona

Pole Winner:
Bobby Isaac, #71 Nord Krauskopf Daytona, 199.658 MPH

Rebel 400
Darlington Raceway (1.366 mi.)
Darlington, SC, May 9, 1970

The Top Five:
1. David Pearson, #17 Holman-Moody Ford
2. Dick Brooks, #32 Dick Brooks SuperBird
3. Bobby Isaac, #71 Nord Krauskopf Daytona
4. James Hylton, #48 Hylton Engineering Ford
5. Benny Parsons, #72 L.G. DeWitt Ford

Other Wing Car Finishes:
15. Buddy Baker, #6 Cotton Owens Daytona
19. Pete Hamilton, #40 Petty Enterprises SuperBird
20. Bobby Allison, #22 Mario Rossi Daytona
22. Charlie Glotzbach, #99 Nichels-Goldsmith Daytona
35. Jim Vandiver, #31 Vandiver Daytona

Pole Winner:
Charlie Glotzbach, #99 Nichels-Goldsmith Daytona, 153.822 MPH

World 600
Charlotte Motor Speedway (1.5 mi.)
Charlotte, NC, May 24, 1970

The Top Five:
1. Donnie Allison, #27 Banjo Matthews Ford
2. Cale Yarborough, #21 Wood Brothers Mercury
3. Benny Parsons, #72 L.G. DeWitt Ford
4. Tiny Lund, #55 McConnell Daytona
5. James Hylton, #48 Hylton Engineering Ford

Other Wing Car Finishes:
6. Bugs Stevens, #36 Richard Brown SuperBird
7. Bobby Isaac, #71 Nord Krauskopf Daytona
8. Pete Hamilton, #40 Petty Enterprises SuperBird
10. Jim Vandiver, #31 Vandiver Daytona
12. Joe Frasson, #18 Frasson Daytona
16. Neil Castles, #06 Castles Daytona
17. Dave Marcis, #30 Marcis Daytona
20. Jim Paschal, #43 Petty Enterprises SuperBird
23. Buddy Baker, #6 Cotton Owens Daytona

24. Fred Lorenzen, #28 Howard Daytona
25. Charlie Glotzbach, #99 Nichels-Goldsmith Daytona
31. Dick Brooks, #32 Dick Brooks SuperBird
39. Bobby Allison, #22 Mario Rossi Daytona

Pole Winner:
Bobby Isaac, #71 Nord Krauskopf Daytona, 159.277 MPH

Motor State 400
Michigan International Speedway (2 mi.)
Brooklyn, MI, June 7, 1970

The Top Five:
1. Cale Yarborough, #21 Wood Brothers Mercury
2. Pete Hamilton, #40 Petty Enterprises SuperBird
3. David Pearson, #17 Holman-Moody Ford
4. LeeRoy Yarbrough, #98 Junior Johnson Ford
5. Bobby Isaac, #71 Nord Krauskopf Daytona

Other Wing Car Finishes:
6. Jim Vandiver, #31 Vandiver Daytona
7. Buddy Baker, #6 Cotton Owens Daytona
9. Dave Marcis, #30 Marcis Daytona
17. Bobby Allison, #22 Mario Rossi Daytona

22. Joe Frasson, #18 Frasson Daytona
24. Charlie Glotzbach, #99 Nichels-Goldsmith Daytona
28. Richard Petty, #43 Petty Enterprises SuperBird
30. Neil Castles, #06 Castles Daytona

Pole Winner:
Pete Hamilton, #40 Petty Enterprises SuperBird, 162.737 MPH

Falstaff 400
Riverside International Raceway (2.62 mi. road course)
Riverside, CA, June 14, 1970

The Top Five:
1. Richard Petty, #43 Petty Enterprises SuperBird
2. Bobby Allison, #22 Mario Rossi Daytona
3. James Hylton, #48 Hylton Engineering Ford
4. John Soares, #08 Soares Plymouth
5. Dick Gulstrand, #44 James Good Chevrolet

Other Wing Car Finishes:
6. Jack McCoy, #7 Ernie Conn Daytona
7. Neil Castles, #06 Castles Daytona
16. Bobby Isaac, #71 Nord Krauskopf Daytona
25. Dick Bown, #02 Mike Ober SuperBird
34. Dave Marcis, #30 Marcis Daytona

Pole Winner:
Bobby Allison, #22 Mario Rossi Daytona, 111.621 MPH

Firecracker 400
Daytona International Speedway (2.5 mi.)
Daytona Beach, FL, July 4, 1970

The Top Five:
1. Donnie Allison, #27 Banjo Matthews Ford
2. Buddy Baker, #6 Cotton Owens Daytona
3. Bobby Allison, #22 Mario Rossi Daytona
4. Charlie Glotzbach, #99 Nichels-Goldsmith Daytona
5. Dick Brooks, #32 Dick Brooks SuperBird

Other Wing Car Finishes:
6. Dr. Don Tarr, #36 Richard Brown SuperBird
9. Bobby Isaac, #71 Nord Krauskopf Daytona
10. Neil Castles, #06 Castles Daytona
18. Richard Petty, #43 Petty Enterprises SuperBird
22. Dave Marcis, #30 Marcis Daytona
23. Joe Frasson, #18 Frasson Daytona
30. Pete Hamilton, #40 Petty Enterprises SuperBird
32. Fred Lorenzen, #28 Ray Fox Daytona
33. Jim Vandiver, #31 O.L. Nixon Daytona

Pole Winner:
Cale Yarborough, #21 Wood Brothers Mercury, 191.640 MPH

Schaefer 300
Trenton Speedway (1.5 mi.)
Trenton, NJ, July 12, 1970

The Top Five:
1. Richard Petty, #43 Petty Enterprises SuperBird
2. Bobby Allison, #22 Mario Rossi Daytona
3. Charlie Glotzbach, #99 Nichels-Goldsmith Daytona
4. Dick Brooks, #32 Dick Brooks SuperBird
5. James Hylton, #48 Hylton Engineering Ford

Other Wing Car Finishes:
19. Bobby Isaac, #71 Nord Krauskopf Daytona
34. Buddy Baker, #6 Cotton Owens Daytona

Pole Winner:
Bobby Isaac, #71 Nord Krauskopf Daytona, 131.749 MPH

Volunteer 500
Bristol International Speedway (.533 mi.)
Bristol, TN, July 19, 1970

The Top Five:
1. Bobby Allison, #22 Allison Dodge

2. LeeRoy Yarbrough, #98 Junior Johnson Ford
3. Bobby Isaac, #71 Nord Krauskopf Dodge
4. G.C. Spencer, #49 Spencer Plymouth
5. Richard Petty, #43 Petty Enterprises Plymouth

Wing Car Finishes:
24. Dick Bown, #02 Mike Ober SuperBird (only wing car entry)

Pole Winner:
Cale Yarborough, #21 Wood Brothers Mercury, 107.375 MPH

East Tennessee 200
Smoky Mountain Raceway (.520 mi.)
Maryville, TN, July 24, 1970

The Top Five:
1. Richard Petty, #43 Petty Enterprises Plymouth
2. Bobby Isaac, #71 Nord Krauskopf Dodge
3. Dick Brooks, #32 Dick Brooks Plymouth
4. James Hylton, #48 Hylton Engineering Ford
5. Friday Hassler, #39 James Hanley Chevrolet

Wing Car Finishes:
6. Dick Bown, #02 Mike Ober SuperBird (only wing car entry)

Pole Winner:
Richard Petty, #43 Petty Enterprises Plymouth, 91.264 MPH

Nashville 420
Fairgrounds Speedway (.596 mi.)
Nashville, TN, July 25, 1970

The Top Five:
1. Bobby Isaac, #71 Nord Krauskopf Dodge
2. Bobby Allison, #22 Allison Dodge
3. Neil Castles, #06 Castles Dodge
4. Cecil Gordon, #24 Gordon Ford
5. J.D. McDuffie, #70 McDuffie Mercury

Wing Car Finishes:
21. Dick Bown, #02 Mike Ober SuperBird (only wing car entry)

Pole Winner:
LeeRoy Yarbrough, #98 Junior Johnson Ford, 114.115 MPH

Dixie 500
Atlanta International Raceway (1.522 mi.)
Hampton, GA, August 2, 1970

The Top Five:
1. Richard Petty, #43 Petty Enterprises SuperBird
2. Cale Yarborough, #21 Wood Brothers Mercury
3. LeeRoy Yarbrough, #98 Junior Johnson Mercury
4. Buddy Baker, #6 Cotton Owens Daytona
5. Donnie Allison, #27 Banjo Matthews Ford

Other Wing Car Finishes:
6. Pete Hamilton, #40 Petty Enterprises SuperBird
7. Bobby Allison, #22 Mario Rossi Daytona
9. Jim Vandiver, #31 Vandiver Daytona
19. Dick Bown, #02 Mike Ober SuperBird
21. Neil Castles, #06 Castles Daytona
22. Fred Lorenzen, #28 Ray Fox Daytona
26. Dick Brooks, #32 Dick Brooks SuperBird
28. Joe Frasson, #18 Frasson Daytona
29. Dave Marcis, #30 Marcis Daytona
33. Charlie Glotzbach, #99 Nichels-Goldsmith Daytona
35. Bobby Isaac, #71 Nord Krauskopf Daytona
40. Bugs Stevens, #36 Richard Brown SuperBird

Pole Winner:
Fred Lorenzen, #28 Ray Fox Daytona, 157.625 MPH

West Virginia 300
International Raceway Park (.437 mi.)
Ona, WV, August 11, 1970

The Top Five:
1. Richard Petty, #43 Petty Enterprises Plymouth
2. James Hylton, #48 Hylton Engineering Ford
3. Neil Castles, #06 Castles Dodge
4. John Sears, #4 Sears Dodge
5. Dave Marcis, #97 Cecil Gordon Ford

Wing Car Finishes:
21. Buddy Baker, #86 Neil Castles Daytona (only wing car entry)

Pole Winner:
LeeRoy Yarbrough, #98 Junior Johnson Ford, 114.115 MPH

Yankee 400
Michigan International Speedway (2 mi.)
Brooklyn, MI, August 16, 1970

The Top Five:
1. Charlie Glotzbach, #99 Nichels-Goldsmith Daytona
2. Bobby Allison, #22 Mario Rossi Daytona
3. Dick Brooks, #32 Dick Brooks SuperBird
4. Bobby Isaac, #71 Nord Krauskopf Daytona
5. Pete Hamilton, #40 Petty Enterprises SuperBird

Other Wing Car Finishes:
6. Buddy Baker, #6 Cotton Owens Daytona
11. Joe Frasson, #18 Frasson Daytona
12. Neil Castles, #06 Castles Daytona

14. Richard Petty, #43 Petty Enterprises SuperBird
39. Dave Marcis, #30 Marcis Daytona

Pole Winner:
Charlie Glotzbach, #99 Nichels-Goldsmith Daytona, 157.363 MPH

Talladega 500
Alabama International Motor Speedway (2.66 mi.)
Talladega, AL, August 23, 1970

Top Five:
1. Pete Hamilton, #40 Petty Enterprises SuperBird
2. Bobby Isaac, #71 Nord Krauskopf Daytona
3. Charlie Glotzbach, #99 Nichels-Goldsmith Daytona
4. David Pearson, #17 Holman-Moody Ford
5. Buddy Baker, #6 Cotton Owens Daytona

Other Wing Car Finishes:
7. Richard Petty, #43 Petty Enterprises SuperBird
8. Ramo Stott, #77 Stott SuperBird
9. Jim Vandiver, #31 O.L. Nixon Daytona
11. Joe Frasson, #18 Frasson Daytona
13. Bobby Allison, #22 Mario Rossi Daytona
32. Dave Marcis, #30 Marcis Daytona
39. Tiny Lund, #55 McConnell Daytona
40. Dick Brooks, #32 Dick Brooks SuperBird
49. Fred Lorenzen, #28 Ray Fox Daytona
50. Neil Castles, #06 Castles Daytona

Pole Winner:
Bobby Isaac, #71 Nord Krauskopf Daytona, 186.834 MPH

Southern 500
Darlington Raceway (1.366 mi.)
Darlington, SC, September 7, 1970

The Top Five:
1. Buddy Baker, #6 Cotton Owens Daytona
2. Bobby Isaac, #71 Nord Krauskopf Daytona
3. Pete Hamilton, #40 Petty Enterprises SuperBird
4. David Pearson, #17 Holman-Moody Ford
5. Richard Petty, #43 Petty Enterprises SuperBird

Other Wing Car Finishes:
6. Charlie Glotzbach, #99 Nichels-Goldsmith Daytona
8. Dick Brooks, #32 Dick Brooks SuperBird
10. Bobby Allison, #22 Mario Rossi Daytona
15. Joe Frasson, #18 Frasson Daytona
28. Neil Castles, #06 Castles Daytona
38. Bugs Stevens, #36 Richard Brown SuperBird

Pole Winner:
David Pearson, #17 Holman-Moody Ford, 150.555 MPH

Mason-Dixon 300
Dover Downs International Speedway (1 mi.)
Dover, DE, September 20, 1970

The Top Five:
1. Richard Petty, #43 Petty Enterprises SuperBird
2. Bobby Allison, #22 Mario Rossi Daytona
3. Charlie Glotzbach, #99 Nichels-Goldsmith Daytona
4. David Pearson, #17 Holman-Moody Ford
5. Benny Parsons, #72 L.G. DeWitt Ford

Other Wing Car Finishes:
6. Bobby Isaac, #71 Nord Krauskopf Daytona
11. Buddy Baker, #6 Cotton Owens Daytona
23. Dave Marcis, #30 Marcis Daytona

Pole Winner:
Bobby Isaac, #71 Nord Krauskopf Daytona, 129.538 MPH

National 500
Charlotte Motor Speedway (1.5 mi.)
Charlotte, NC, October 11, 1970

The Top Five:
1. LeeRoy Yarbrough, #98 Junior Johnson Mercury
2. Bobby Allison, #22 Mario Rossi Daytona
3. Fred Lorenzen, #3 Ray Fox Daytona
4. Benny Parsons, #72 L.G. DeWitt Ford
5. Bobby Isaac, #71 Nord Krauskopf Daytona

Other Wing Car Finishes:
23. Richard Petty, #43 Petty Enterprises SuperBird
24. Pete Hamilton, #40 Petty Enterprises SuperBird
25. Butch Hirst, #36 Richard Brown SuperBird
26. Buddy Baker, #6 Cotton Owens Daytona
27. Neil Castles, #06 Castles Daytona
30. Dick Brooks, #32 Dick Brooks SuperBird
32. Marty Robbins, #42 Robbins Daytona
33. Jim Vandiver, #31 O.L. Nixon Daytona
36. Dave Marcis, #30 Marcis Daytona
37. Charlie Glotzbach, #99 Nichels-Goldsmith Daytona

Pole Winner:
Charlie Glotzbach, #99 Nichels-Goldsmith Daytona, 157.273 MPH

American 500
North Carolina Motor Speedway (1.017 mi.)
Rockingham, NC, November 15, 1970

The Top Five:
1. Cale Yarborough, #21 Wood Brothers Mercury
2. David Pearson, #17 Holman-Moody Ford
3. Bobby Allison, #22 Mario Rossi Daytona
4. Donnie Allison, #27 Banjo Matthews Ford
5. Buddy Baker, #6 Cotton Owens Daytona

Other Wing Car Finishes:
6. Richard Petty, #43 Petty Enterprises SuperBird
7. Bobby Isaac, #71 Nord Krauskopf Daytona
11. Joe Frasson, #18 Frasson Daytona
12. Neil Castles, #06 Castles Daytona
15. Pete Hamilton, #40 Petty Enterprises SuperBird
16. Jim Vandiver, #31 Vandiver Daytona
23. Roy Mayne, #36 Richard Brown SuperBird
31. Charlie Glotzbach, #99 Nichels-Goldsmith Daytona
32. Dave Marcis, #30 Marcis Daytona
35. Tiny Lund, #55 McConnell Daytona
39. Dick Brooks, #32 Dick Brooks SuperBird

Pole Winner:
Charlie Glotzbach, #99 Nichels-Goldsmith Daytona, 136.498 MPH

125-Mile Qualifying Race Two for 1971 Daytona 500
Daytona International Speedway (2.5 mi.)
Daytona Beach, FL, February 11, 1971

The Top Five:
1. David Pearson, #17 Holman-Moody Mercury
2. Buddy Baker, #11 Petty Enterprises Dodge
3. Dick Brooks, #22 Mario Rossi Daytona (only wing car entry)
4. Bill Dennis, #90 Junie Donleavey Mercury
5. Benny Parsons, #72 L.G. DeWitt Ford

Pole Winner:
Bobby Isaac, #71 Nord Krauskopf Dodge, 180.050 MPH

Daytona 500
Daytona International Speedway (2.5 mi.)
Daytona Beach, FL, February 14, 1971

The Top Five:
1. Richard Petty, #43 Petty Enterprises Plymouth
2. Buddy Baker, #11 Petty Enterprises Dodge
3. A.J. Foyt, #21 Wood Brothers Mercury
4. David Pearson, #17 Holman-Moody Mercury
5. Fred Lorenzen, #99 Nichels-Goldsmith Plymouth

Wing Car Finishes:
7. Dick Brooks, #22 Mario Rossi Daytona (only wing car entry)

Pole Winner:
A.J. Foyt, #21 Wood Brothers Mercury, 182.744 MPH

UNITED STATES AUTO CLUB WING CAR COMPETITION RESULTS

In USAC competition, the wing cars generally made appearances at larger tracks such as Michigan International Speedway, Pocono International Raceway, and the famous "Milwaukee Mile" track in West Allis, WI. Here are the results by race winner and wing car placings.

April 12, 1970; 200 miles at Sears Point, CA (2.5 mi. road course)
1. Roger McCluskey SuperBird

April 18, 1970; 200 miles at Phoenix, AZ (1 mi. oval)
1. A.J. Foyt Ford
2. Roger McCluskey SuperBird
6. Ray Elder Daytona
7. Terry Nichels Daytona

May 3, 1970; 250 miles at Clermont, IN (2.5 mi. road course)
1. A.J. Foyt Ford
3. Norm Nelson SuperBird
5. Terry Nichels Daytona
11. Butch Hartman Daytona

July 4, 1970; 200 miles at Brooklyn, MI (2 mi. oval)
1. A.J. Foyt Ford
2. Don White Daytona
3. Roger McCluskey SuperBird
4. Sal Tovella SuperBird

July 12, 1970; 200 miles at Milwaukee, WI (1 mi. oval)
1. Roger McCluskey SuperBird
3. Sal Tovella SuperBird
4. Butch Hartman Daytona

July 19, 1970; 200 miles at Dover, DE (1 mi. oval)
1. A.J. Foyt Ford
2. Don White Daytona
3. Roger McCluskey SuperBird
4. Butch Hartman Daytona

August 16, 1970; 150 miles at Milwaukee, WI (1 mi. oval)
1. Don White Daytona
2. Roger McCluskey SuperBird
9. Sal Tovella SuperBird

August 20, 1970; 200 miles at Milwaukee, WI (1 mi. oval)
1. A.J. Foyt Ford
3. Roger McCluskey SuperBird
5. Sal Tovella SuperBird

September 20, 1970; 250 miles at Milwaukee, WI (1 mi. oval)
1. Jack Bowsher Ford
2. Roger McCluskey SuperBird
4. Lem Blankenship SuperBird
5. Butch Hartman Daytona

July 11, 1971; 200 miles at Milwaukee, WI (1 mi. oval)
1. Jack Bowsher Ford
3. Roger McCluskey SuperBird
4. Butch Hartman Daytona
5. Dave Whitcomb Daytona
24. Norm Nelson SuperBird
30. Sal Tovella SuperBird
32. Verlin Eaker Daytona
36. Don White Daytona

July 18, 1971; 200 miles at Brooklyn, MI (2 mi. oval)
1. Roger McCluskey SuperBird
4. Lem Blankenship SuperBird
5. Butch Hartman Daytona
8. Don White Daytona
10. Sal Tovella SuperBird
24. Verlin Eaker Daytona

August 19, 1971; 150 miles at Milwaukee, WI (1 mi. oval)
1. A.J. Foyt Ford
2. Butch Hartman Daytona
3. Dave Whitcomb Daytona
4. Sal Tovella SuperBird
5. Verlin Eaker Daytona
12. Norm Nelson SuperBird
28. Roger McCluskey SuperBird
31. Don White Daytona

August 22, 1971; 200 miles at Milwaukee, WI (1 mi. oval)
1. Roger McCluskey SuperBird
2. Lem Blankenship SuperBird
3. Norm Nelson SuperBird
4. Verlin Eaker Daytona
5. Sal Tovella SuperBird
11. Butch Hartman Daytona

September 12, 1971; 250 miles at Milwaukee, WI (1 mi. oval)
1. Al Unser Ford
2. Roger McCluskey SuperBird
4. Norm Nelson SuperBird
5. Verlin Eaker Daytona
6. Butch Hartman Daytona
7. Lem Blankenship SuperBird
11. Don White Daytona
13. Sal Tovella SuperBird
35. Dave Whitcomb Daytona

September 19-25, 1971 (rain); 500 miles at Pocono, PA (2.5 mi. tri-oval)
1. Butch Hartman Daytona
3. Don White Daytona
5. Lem Blankenship SuperBird
7. Dave Whitcomb Daytona
10. Dick Tobias Daytona
12. Earl Wagner/Geoff Bodine Daytona
15. Gary Bettenhausen Daytona
19. Roger McCluskey SuperBird
27. Verlin Eaker Daytona
34. Norm Nelson SuperBird

July 9, 1972; 200 miles at Milwaukee, WI (1 mi. oval)
1. Roger McCluskey SuperBird
31. Sal Tovella SuperBird
34. Bobby Unser SuperBird

July 16, 1972; 200 miles at Brooklyn, MI (2 mi. oval)
1. Ramo Stott SuperBird
2. Roger McCluskey SuperBird
6. Sal Tovella SuperBird

July 30, 1972; 500 miles at Pocono, PA (2.5 mi. tri-oval)
1. Roger McCluskey SuperBird
11. Sal Tovella SuperBird
21. Ramo Stott SuperBird

August 17, 1972; 150 miles at Milwaukee, WI (1 mi. oval)
1. Gordon Johncock Chevrolet
2. Bobby Unser SuperBird
6. Ramo Stott SuperBird
33. Sal Tovella SuperBird

August 20, 1972; 200 miles at Milwaukee, WI (1 mi. oval)
1. Jack Bowsher Ford
2. Bobby Unser SuperBird
5. Ramo Stott SuperBird
9. Roger McCluskey SuperBird
19. Sal Tovella SuperBird

September 10, 1972; 250 miles at Milwaukee, WI (1 mi. oval)
1. Don White Dodge
5. Roger McCluskey SuperBird
6. Sal Tovella SuperBird
30. Ramo Stott SuperBird
39. Bobby Unser SuperBird

AUTOMOBILE RACING CLUB OF AMERICA WING CAR COMPETITION RESULTS

The presence of wing cars was a rarity in ARCA competition as the majority of ARCA events during the early 1970s took place on smaller tracks in the Midwest. However, during ARCA events held at Daytona and Talladega, the wing cars ruled the roost much as they did in NASCAR competition. The ARCA 300 on February 15, 1970, at Daytona International Speedway found Ramo Stott in the number 7 SuperBird battling down the backstretch with Bobby Watson in the number 8 Charger Daytona. Stott narrowly took the victory. A 50-mile qualifying race at Talladega for the June ARCA event was held on April 11, 1970, and again Stott beat Watson to the line. When the Vulcan 500 was run at Alabama International Motor Speedway on June 14, 1970, Stott and Watson started the race side-by-side as a result of the April qualifier. Ramo led 154 of the 188 laps in his SuperBird, and went on to claim victory in the final superspeedway race of the 1970 ARCA season. Stott ran the entire 500-mile race on one set of Goodyear tires.

WING CAR SPEED RECORDS

On March 24, 1970, Larry Rathgeb and the Chrysler team arrived at Alabama International Motor Speedway with the number 88 engineering Charger Daytona and driver Buddy Baker. Their goal was to see the Daytona become the first car to officially break the 200 MPH mark on a closed course. Just after 4:25 p.m., on the 30th lap of the session, Baker blasted into record territory with a lap of 200.096 MPH around the 2.5-mile speedway. Later that afternoon the number 88 car ran even faster, wrapping things up with a record 200.44795 MPH lap.

Although he won the 1970 NASCAR Grand National championship, there was still one thing Bobby Isaac wanted to do as the year drew to a close — break Buddy Baker's record. On November 24, 1970, the K&K Insurance Daytona began its speed quest at Alabama International Motor Speedway, Bobby Isaac's 22nd lap of the day was a new record — 201.104 MPH.

Bobby Isaac, Harry Hyde, and the K&K Insurance Charger Daytona team then moved on to challenge the Bonneville salt flats from September 13-16, 1971. They set a host of new speed records, obliterating some of the previous marks by more than 25 MPH. Among the USAC-sanctioned records set were the following:

Standing-start, 10 kilometers: 172.483 MPH
Standing-start, 10 miles: 182.174 MPH
100 kilometers: 193.168 MPH
100 miles: 194.290 MPH
Flying mile: 216.946 MPH
Flying kilometer: 217.368 MPH

WHERE THE WINGS ARE

Street versions of the Dodge Charger Daytona and Plymouth SuperBird can be seen at regional Chrysler car gatherings, or at one of the larger annual gatherings such as the Chryslers at Carlisle event in Pennsylvania. These events are often listed in magazines that feature Chrysler products or muscle cars in general. Wing car followers have formed two organizations dedicated to Daytonas and SuperBirds:

Daytona-SuperBird Auto Club	Wing Warriors
3717 Green Meadow Drive	216 12th Street
New Berlin, WI 53151	Boone, IA 50036

Some of the greatest competition wing cars can be seen at racing museums. The International Motorsports Hall of Fame and Museum located at Alabama's Talladega Superspeedway (205-362-5002) contains a fantastic collection of all types of racing machinery, including the number 88 Dodge Charger Daytona with which Buddy Baker shattered the 200 MPH mark and the number 71 Daytona the late Bobby Isaac drove to the 1970 Grand National Championship. At South Carolina's Darlington Raceway you'll find the National Motorsports Press Association Hall of Fame and Joe Weatherly Museum (803-393-2103), where Buddy Baker's famous number 6 Charger Daytona now rests amidst another exciting collection of race cars. Finally, one of Richard Petty's long and low Plymouth SuperBird stock cars can be found in the impressive Richard Petty Museum (919-495-1143) in Level Cross, North Carolina.